John Millar

Elements of descriptive geometry

John Millar

Elements of descriptive geometry

ISBN/EAN: 9783337278489

Printed in Europe, USA, Canada, Australia, Japan

Cover: Foto ©berggeist007 / pixelio.de

More available books at **www.hansebooks.com**

ELEMENTS

OF

DESCRIPTIVE GEOMETRY.

BY

J. B. MILLAR, M.E.

CIVIL ENGINEER; LECTURER ON ENGINEERING IN THE
VICTORIA UNIVERSITY, MANCHESTER.

SECOND EDITION.

London:
MACMILLAN AND CO.
AND NEW YORK
1887

PREFACE.

I HAVE endeavoured, by putting the subject in a simple, concise, and systematic form, to give to this treatise the elementary character which is required in a book intended for beginners, and at the same time to make it sufficiently comprehensive to meet the wants of a more advanced class of students.

The difficulties which hinder beginners I have found to be chiefly of two kinds. One of these arises from the want of sufficient knowledge of solid geometry; the study of projections, as a practical subject, is begun too commonly before the student has made himself acquainted with the geometrical principles on which the solutions of the problems depend. To begin in that way is, I think, to make a mistake; for, without a knowledge of first principles, it is impossible to get such a grasp of the subject as will make it the useful and effective instrument which it ought to be. I have, therefore, considered it best to devote the first chapter to some theorems on the straight line and plane, and to introduce occasional theorems in the other parts of the work; my object being to establish the principles before giving their applications.

The other difficulty to which I have alluded lies in the inability of the learner to realise from their projections

the positions of points and lines in space. It is one which
requires considerable time and thought to overcome. I have
tried, however, to reduce it as much as possible by giving
two figures with each problem of Chapter II.; one of these
figures is a perspective, representing the points, lines, and
planes in their true positions, and the other shows their
projections, and the ordinary solution of the problem. I
trust that by carefully comparing these figures the student
may be led by easy steps to connect the two things and
obtain a clear idea of the methods employed in Descriptive Geometry. I have little doubt that any one who
masters the first two chapters will find his after-course both
interesting and comparatively easy.

I may add that I have never lost sight of the practical
nature of the subject, and have introduced only so much
theory as seemed to me necessary to place the practice on a
proper footing.

J. B. M.

CONTENTS.

CHAPTER I.

THE STRAIGHT LINE AND PLANE.

CHAPTER II.

INTRODUCTION TO DESCRIPTIVE GEOMETRY, AND PROBLEMS ON THE STRAIGHT LINE AND PLANE.

CONTENTS.

PAGE

CHAPTER III.

EXAMPLES OF THE PROJECTIONS OF PLANE AND SOLID FIGURES. SOLUTION OF THE SPHERICAL TRIANGLE.

CHAPTER IV.

CURVED SURFACES AND TANGENT PLANES.

xii

CONTENTS.

PAGE

Theorem II. The tangent plane to a cone at any point is a tangent at every point of the generator passing through that point . . 112

Problem II. To draw a tangent plane to a cone through a given point on the surface 114

Problem III. To draw a tangent plane to a cone through a given external point 116

Problem IV. To draw a tangent plane to a cone which shall be parallel to a given straight line 118

Problem V. To find the traces of a plane which shall contain a given line and have a given inclination 120

Problem VI. To find the traces of a plane which shall contain a given line, and make a given angle with a given plane 122

Problem VII. To find the section of a cone of revolution by a plane . 124

Problem VIII. To find the development of a given conical surface . 130

THE CYLINDER 130

Problem IX. Given one projection of a point on a cylinder of revolution, to find the other projection, when the axis of the cylinder is parallel to the ground-line 134

Problem X. Through a given point on a cylinder of revolution to draw a tangent plane to the surface, when the axis of the cylinder is parallel to the ground-line 136

Problem XI. Through a given external point to draw a tangent plane to a cylinder of revolution, when the axis is parallel to the ground-line 136

Problem XII. To find the development of a right circular cylinder . 138

SURFACES OF REVOLUTION 140

Problem XIII. Given one projection of a point on a surface of revolution, to determine the other projection 142

Problem XIV. To draw a tangent plane to a surface of revolution at a given point on the surface 142

THE SPHERE 144

Theorem III. Every plane section of a sphere is a circle . . . 145

Problem XV. Given the projections of four points on the surface of a sphere, to find its centre and radius 146

CHAPTER V.

INTERSECTIONS OF CURVED SURFACES.

CHAPTER VI.

AXOMETRIC PROJECTION.

ELEMENTS OF DESCRIPTIVE GEOMETRY.

CHAPTER I.

THEOREMS.

THE STRAIGHT LINE AND PLANE.

DEFINITION 1. A *figure* which has length, breadth, and thickness, is called a *solid*.

DEF. 2. A *surface* is the boundary of a solid and has length and breadth only.

DEF. 3. A *plane* is a *surface* such that if any two points be taken in it, the straight line passing through them lies wholly in that surface.

The plane is said to contain the line.

THEOREM I.

Two straight lines which cut one another are in one plane.

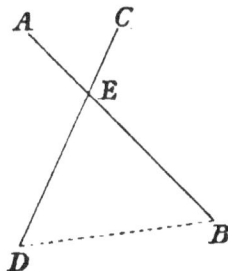

Let the two straight lines AB, CD, intersect in the point E, then AB and CD are contained by one plane.

Proof. Let any plane which contains *AB* be made to revolve about that line as an axis, there is one position in which it will contain the point *D*; but it also contains the point *E*; therefore it contains the whole line *CD*, ... (Def. 3).

Therefore *AB* and *CD* are in one plane.

Corollary. The line joining any two points *B* and *D*, one on each line, will be in the plane containing *AB*, *CD*,...... (Def. 3).

Therefore three straight lines which meet one another, not in the same point, are in one plane.

THEOREM II.

If two planes cut one another, their common section is a straight line.

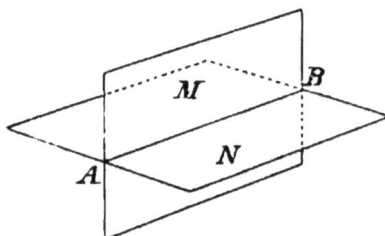

Let *M* and *N* be the two intersecting planes, their com-on section is the straight line *AB*.

Proof. Let *A* and *B* be two points common to both nes *M* and *N*.

Since the points *A* and *B* are in the plane *M*, the straight joining them lies wholly in that plane, ... (Def. 3).

Similarly, since *A* and *B* are in the plane *N*, the straight *AB* lies wholly in *N*.

Then the straight line *AB* is the common section of ˀd *N*.

ᵥ. 4. The inclination to one another of two lines

which do not meet is the angle contained by two intersecting lines parallel to them, each to each.

DEF. 5. A straight line is perpendicular to a plane when it is at right angles to every line meeting it in that plane.

The *foot* of the perpendicular is the point in which it meets the plane, and the line is called the *normal* to the plane at that point.

DEF. 6. The angle between two intersecting planes is called a dihedral angle, and is measured by the angle between two straight lines drawn from any point of their common section, at right angles to it, one in each plane.

DEF. 7. When the angle between two planes is a right angle, the planes are said to be perpendicular to one another.

THEOREM III.

If a straight line be perpendicular to each of two straight lines at their point of intersection, it shall also be perpendicular to their plane.

Let *AD* be perpendicular to *DB* and *DC* at their point of intersection *D*; it is required to prove that *AD* is perpendicular to the plane *BDC*.

1—2

Let AD be produced to E, so that $DE = DA$, and E joined with B, C and F; DF being any line whatever in the plane BDC, and F on the straight line BC.

Proof. In the two triangles ADB, EDB, $AD = DE$ by construction, BD is common to both triangles, and the angle $ADB = BDE$, since each is a right angle.

Therefore $AB = BE$.

It may be proved in a similar way that $AC = CE$. Therefore the two triangles ABC, EBC have the three sides of the one respectively equal to the three sides of the other, and are consequently equal in every respect. Hence if the triangle EBC were turned about BC till the planes of the two triangles coincided, E would coincide with A and EF with AF.

Therefore $AF = EF$.

Then in the triangles ADF, EDF the three sides of the one are respectively equal to the three sides of the other.

Therefore the angle $ADF = EDF$.

That is ADF and EDF are right angles.

Therefore AD is perpendicular to DF.

In a similar way it may be shown that AD is perpendicular to every line passing through D in the plane BDC, and is therefore perpendicular to the plane.

Cor. 1. It follows from Def. 4 that if AD is perpendicular to any two lines in the plane it is perpendicular to every line in it.

Cor. 2. Any number of straight lines which are drawn at right angles to the same straight line from the same point of it, are all in the plane which is perpendicular to the line at that point.

Hence if one of the arms of a right angle be made to revolve about the other as an axis, it will describe a plane normal to that axis.

Cor. 3. Through any given point, either within or without a plane, only one normal can be drawn to the plane. For if it were possible to draw more than one, each of them would be perpendicular to a straight line in the same plane with them, which is impossible.

THEOREM IV.

Every plane which contains the normal to another plane is perpendicular to that plane.

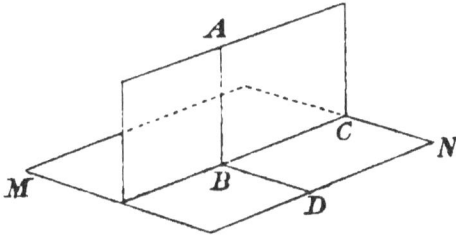

Let AB be normal to the plane MN; it is required to prove that any plane ABC which contains AB is perpendicular to MN.

Proof. Let BD in MN be a perpendicular to BC, the common section of the two planes.

Because AB is perpendicular to the plane MN it is perpendicular to BC and BD,......(Def. 5).

But the angle between the planes is measured by the angle ABD,......(Def. 6).

Therefore the plane ABC is perpendicular to MN,...... (Def. 7).

THEOREM V.

If two planes be perpendicular to one another, every line drawn in one of them perpendicular to their common section shall be perpendicular to the other.

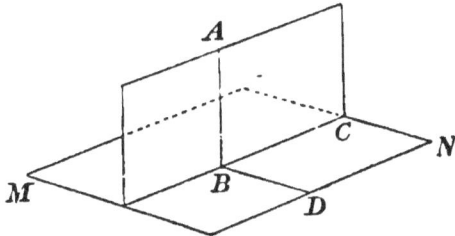

Let the plane AC be perpendicular to the plane MN, and

the line AB in AC perpendicular to BC, the common section of the two planes; it is required to prove that AB is a normal to the plane MN.

Proof. Let BD be a perpendicular to BC in the plane MN.

Then because the plane AC is perpendicular to MN, the angle ABD is a right angle,......(Def. 7).

Therefore AB is perpendicular to BD and BC, and consequently to the plane DC,......(Theor. III.).

Cor. If from any point of the plane AC a normal be drawn to MN, that normal must lie wholly in AC, for if not, two normals could be drawn to MN from the same point in AC, which is impossible,......(Theor. III. Cor. 3).

THEOREM VI.

. *If two planes which cut one another be both perpendicular to a third plane, their common section shall be perpendicular to the same plane.*

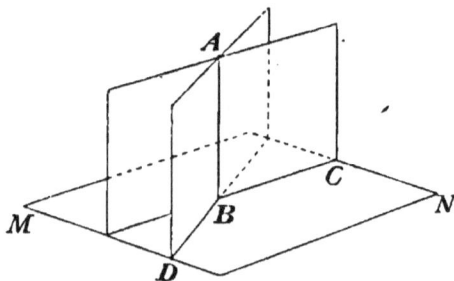

Let the planes AC and AD be each perpendicular to MN; it is required to prove that their common section AB is perpendicular to MN.

Proof. Let BC and BD be the lines of intersection of the planes AC and AD with MN.

The line drawn through B perpendicular to the plane MN must lie wholly in the plane AC,......(Theor. v. Cor.) Similarly it must lie in AD.

Therefore AB is the normal to MN at the point B.

THEOREM VII.

Two straight lines which are perpendicular to the same plane are parallel to one another.

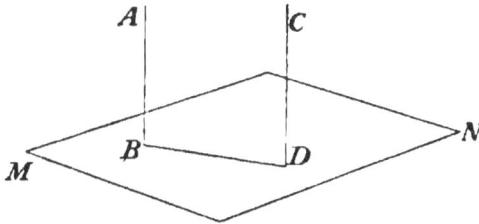

Let the straight lines AB and CD be each perpendicular to MN; it is required to prove that they are parallel to one another.

Proof. Let B and D be the points of intersection of the lines with MN.

Because AB is perpendicular to MN the plane ABD is perpendicular to it also,......(Theor. IV.).

Because the plane ABD is perpendicular to MN, and the line DC is drawn through D perpendicular to MN, it lies wholly in the plane ABD,......(Theor. v. Cor.).

Therefore AB and CD are in the same plane.

Also, since AB and CD are each perpendicular to MN, BD is their common perpendicular,......(Def. 5).

Hence AB and CD are in the same plane, and the straight line BD, cutting them, makes the angles B and D two right angles.

Therefore the lines are parallel.

THEOREM VIII.

If two straight lines be parallel, and one of them be perpendicular to a plane, the other shall be perpendicular to it also.

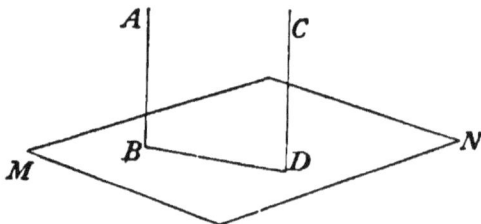

Let *AB* and *CD* be parallel, and *AB* perpendicular to *MN*; it is required to prove that *CD* is also perpendicular to *MN*.

Proof. Let *B* and *D* be the points of intersection of *AB* and *CD* with *MN*.

AB and *CD* are in the same plane, being parallel, and *BD* is in the same plane with them,......(Def. 3).

AB and *CD* being parallel, the angles *ABD* and *CDB* are together equal to two right angles; but since *AB* is perpendicular to *MN*, *ABD* is a right angle; therefore *CDB* is a right angle.

Because *AB* is perpendicular to *MN*, the plane *ABD* is also perpendicular to *MN*,......(Theor. IV.).

But it has been proved that *CD* is in the plane *ABD*, and that it is perpendicular to *BD*, the common section of the two planes.

Therefore *CD* is perpendicular to *MN*,......(Theor. V.).

THEOREM IX.

Two straight lines which are each parallel to the same straight line are parallel to one another.

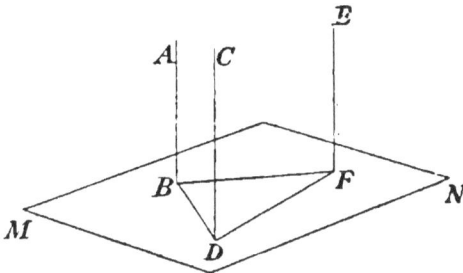

Let CD and EF be each parallel to AB; it is required to prove that they are parallel to one another.

Proof. Let the plane MN be perpendicular to AB.

Then CD and EF are each perpendicular to MN, ...
(Theor. VIII.).

And consequently parallel to one another, ... (Theor. VII.).

DEF. 8. A straight line and a plane are *parallel* to one another when they cannot meet, however far they may be produced.

DEF. 9. Planes which do not meet when indefinitely produced in every direction are *parallel* to one another.

It follows from definitions 8 and 9 that if two planes are parallel to one another, any line in one of them must be parallel to the other.

THEOREM X.

If two straight lines are parallel to one another, any plane which contains one of them, but not both, is parallel to the other.

Let AB and CD be parallel; it is required to prove that AB is parallel to the plane MN which contains CD, but not AB.

Proof. Because AB and CD are parallel they are in the same plane $ABDC$.

Therefore if AB meet the plane MN it must meet it in some point of the line CD produced, which is the common section of the two planes.

But AB cannot meet CD, being parallel to it. Therefore it cannot meet the plane MN.

Therefore AB is parallel to MN,......(Def. 8).

THEOREM XI.

If a straight line be parallel to a plane, it shall be parallel to the line in which any plane containing it cuts the first plane.

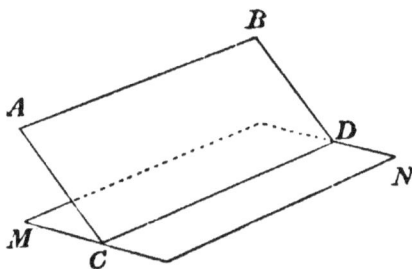

Let AB be parallel to the plane MN, and let the plane $ABDC$ cut MN in CD; it is required to prove that AB is parallel to CD.

Proof. If AB be not parallel to CD, it must meet it if produced, since the two lines are in the same plane. But in the same point it would also meet the plane MN, which is impossible, since it is parallel to it.

Therefore AB cannot meet CD; and being in the same plane the lines are parallel.

Cor. 1. If two straight lines which meet be each parallel to the same plane, the plane containing the two lines shall be parallel to that plane. For if the two planes were not parallel they would meet when produced, and their common section would be parallel to each of two intersecting lines, which is impossible.

Cor. 2. If two straight lines AB, CD are parallel to one another, they are each parallel to EF, the common section of two planes $ABFE$, $CDFE$, each of which contains one of the lines. For AB is parallel to the plane $CDFE$, and CD parallel to the plane $ABFE$, ... (Theor. X.)

Theorem XII.

If two parallel planes be cut by another plane, their common sections with it shall be parallel.

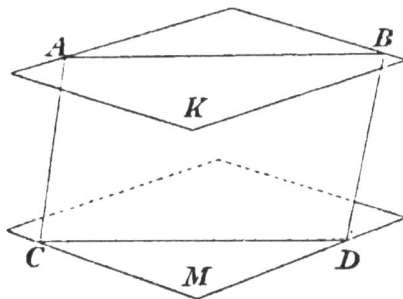

Let the parallel planes K and M be cut by the plane ABD in the lines AB, CD, respectively; it is required to prove that AB and CD are parallel.

Proof. Because AB and CD lie wholly in the planes K and M respectively, they cannot meet except the planes also meet one another.

But the planes cannot meet, since they are parallel. Therefore AB and CD cannot meet; and as they are in the same plane, $ABDC$, they must be parallel.

Theorem XIII.

Planes to which the same straight line is perpendicular are parallel to one another.

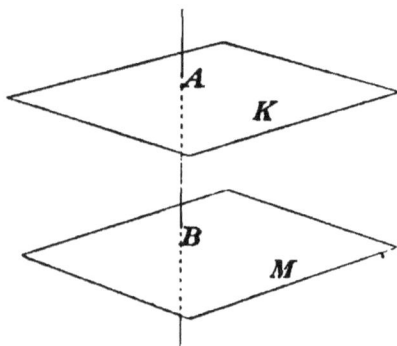

Let the line AB be perpendicular to the planes K and M, intersecting them at the points A and B respectively; it is required to prove that K and M are parallel to one another.

Proof. The two planes cannot meet, for if they met, two lines might be drawn from a point of their common section, one in each plane, to the points A and B, and the line AB would be at right angles to both lines, ...(Def. 5).

That is, if the planes met, two perpendiculars could be drawn to a line from the same point without it, which is impossible. Therefore the planes are parallel.

Theorem XIV.

If two straight lines, which meet one another, be parallel respectively to two other lines which meet, but are not in the same plane with the first two, the first two and the other two shall contain equal angles, and their planes shall be parallel.

Let AB, BC be parallel respectively to DE, EF; it is required to prove that the angle $ABC = DEF$, and that the plane ABC is parallel to DEF.

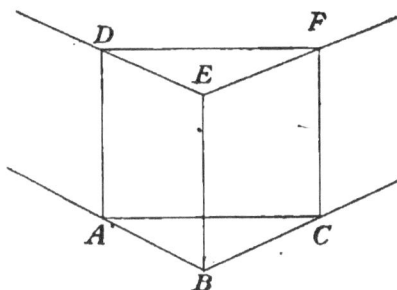

Proof. Let equal distances AB, BC, ED, EF be set off on the four lines, and let A and D be joined with C and F respectively; also A, B and C with D, E and F respectively.

Then, because AB and DE are equal and parallel, AD and BE are also equal and parallel.

Similarly CF and BE are equal and parallel.

Therefore AD and CF are equal and parallel,......(Theor. IX.).

And consequently $AC = DF$.

As the two triangles ABC, DEF have the three sides of the one respectively equal to the three sides of the other the angle $ABC = DEF$.

Again, because AB is parallel to DE, it is also parallel to the plane DEF,......(Theor. X.).

Similarly BC is parallel to DEF.

Therefore the plane ABC is parallel to the plane DEF,... (Theor. XI. Cor. 1).

DEF. 10. The foot of the perpendicular drawn from a point to a plane is called the *orthogonal projection* of the point on the plane.

When not otherwise specified, the word *projection* is understood to mean the orthogonal projection.

DEF. 11. The *projection of a line on a plane* is the line which contains the projections of all points of the given line.

DEF. 12. The perpendicular drawn from the point to the plane is called the *projector* of the point, and the surface which contains the projectors of all the points of a line is called the *projecting surface* of that line. If the projecting surface be a plane it is called the *projecting plane*.

THEOREM XV.

The *projection of a straight line on a plane is a straight line.*

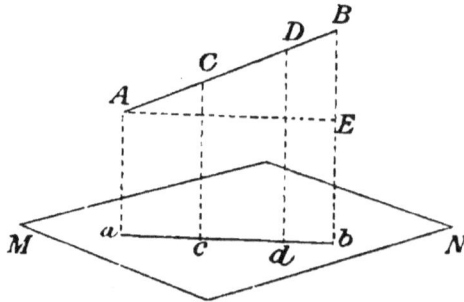

Let AB be a straight line; it is required to prove that its projection on the plane MN is also a straight line.

Proof. Let ABb be a plane containing AB and perpendicular to MN.

Then every line drawn from AB perpendicular to MN must lie in the plane ABb,...(Theor. V. Cor.).

That is, the *projector* of every point of AB lies in the plane ABb, which is, therefore, the projecting plane of AB, and the projection of AB is the common section of the two planes, or the straight line ab.

Cor. 1. If a straight line be parallel to a plane, it shall be equal to its projection on that plane, for the two lines are the opposite sides of a rectangle.

Cor. 2. Since Aa, Cc, Dd are parallel to one another (Theor. VII.), $AC : CD :: ac : cd$.

Therefore, if a straight line be divided into any number of parts, they will have the same ratio to one another that their projections have.

THEOREM XVI.

If two straight lines be parallel to one another, their projections on the same plane shall also be parallel.

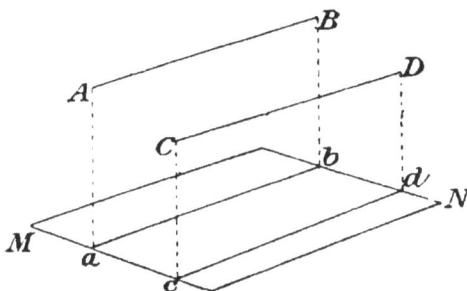

Let AB and CD be parallel; it is required to prove that their projections ab, cd on MN are parallel.

Proof. Because Aa and Cc are both perpendicular to the same plane MN (Def. 10), they are parallel to one another, ...(Theor. VII.).

Therefore, the plane aAB is parallel to cCD,...(Theor. XIV.).

But ab, cd are the common sections of these two parallel planes with MN.

Therefore ab is parallel to cd,......(Theor. XII.).

THEOREM XVII.

If two straight lines be at right angles to one another, their projections on a plane parallel to any one of them shall also be at right angles.

Let the lines AB, CD be at right angles to one another, and the plane MN parallel to AB; it is required to prove that the projections ab, cd are also at right angles.

Proof. Because AE is parallel to the plane MN it is parallel to ae,......(Theor. XI.).

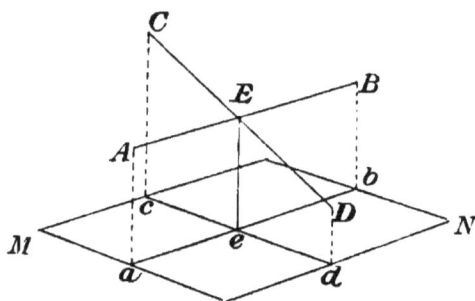

And since Ee is at right angles to ae, it is also at right angles to AE.

Therefore AE is perpendicular to the plane CDd, since it is perpendicular to the two lines CD, Ee,......(Theor. III.).

But ae being parallel to AE is also perpendicular to the plane CDd, and consequently to the line cd. That is, the angle aed between the projections of AB, CD is a right angle.

THEOREM XVIII.

If the projections of two straight lines be at right angles to one another, and one of the lines parallel to the plane of projection, the two lines shall be at right angles to one another.

If ab and cd be at right angles and AB be parallel to the plane MN, then AB and CD are at right angles.

Proof. Since ae is at right angles to Ee and ed, it is a normal to the plane CDd.

But AE is parallel to MN, and therefore also to ae,...... (Theor. XI.).

Hence AE is normal to CDd,......(Theor. VIII.).

Therefore AB is at right angles to CD,......(Def. 5).

CHAPTER II.

DESCRIPTIVE GEOMETRY has for its principal objects the representation of solid figures on a plane surface, and the *graphic* solution of the problems of Solid Geometry.

In other words, it is that branch of Geometry by means of which accurate drawings of machines and structures are made, and problems respecting solid figures reduced to those of Plane Geometry.

As an illustration of the use and importance of this branch of Geometry, suppose the position of a plane and point in space to be known, and that it is required to find the distance of the point from the plane. This will require the drawing of a perpendicular from the point to the plane, finding the point of intersection, and then determining the distance between the two points. Euclid shows how a perpendicular from a point to a plane may be drawn, but in solving the problem practically according to his directions it would be found necessary to work on three different planes, the relative positions of which cannot be determined beforehand, but must be fixed at the different steps of the construction. In attempting to solve this elementary problem by Euclid's method, the necessity will soon be felt of adopting some mode of construction in which the data of the problem may be represented and the necessary con-

M. 2

structions executed on a single plane surface. Both these advantages are obtained by the method of projections (see defs. 10 to 12), which is the method used in Descriptive Geometry.

A point is completely determined when its projections are given on two intersecting planes, the positions of which are known; for there will then be two lines—the projectors—on which the point must lie, and it will consequently be their point of intersection. Let a and a' be the projections of a point on the two planes II and V (fig. 1), then the point must be in each of the lines Aa and Aa'; so that there is only one point A which can have the projections a and a'.

A line, straight or curved, is completely determined when its projections are given on two intersecting planes; as will be seen from fig. 3, where in one case the line AB is the common section of the two projecting planes $ABab$, $ABa'b'$, and in the other the line $BCDE$ is the common section of the two projecting surfaces (def. 12) $BCDEedcb, BCDEe'd'c'b'$. The only exception would be when the two projecting planes of the line coincided, that is when they were perpendicular to the common section of the two fixed planes (Theorem VI.), and in that case it would be necessary to have a projection of the line on another plane not parallel to either of the first two.

These fixed planes to which points and lines are referred are called *co-ordinate planes* or *planes of projection*. They are taken at right angles to one another and one of them is supposed to be horizontal and the other vertical; thus (H) in fig. 1 is *the horizontal plane of projection*, and (V) *the vertical plane of projection*. Their common section xy is called *the ground line*.

Projections on the horizontal plane are called *horizontal projections*, and those on the vertical plane *vertical projections*. The equivalent terms *plan* and *elevation* have been long in use in connection with the drawings of buildings and other objects, and are now frequently extended to the projections of points and lines.

As the planes are at right angles to one another any point on one of them has for its projection on the other a point on the ground line (Theor. V.); thus in fig. 9, a' and b, on the ground line, are the projections of the points A and B respectively, one in each plane of projection.

The plane which contains the two projectors of a point, as aAa', fig. 1, must be perpendicular to each plane of projection (Theor. IV.), and consequently to the ground line (Theor. VI.), so that xy is at right angles to aa_0 and $a'a_0$: hence—

The perpendiculars drawn from the projections of a point to the ground line meet it in the same point.

The lines aa_0 and $a'a_0$ are respectively equal to the projectors Aa' and Aa; *that is aa_0 and $a'a_0$ are equal to the distances of A from the planes of projection.*

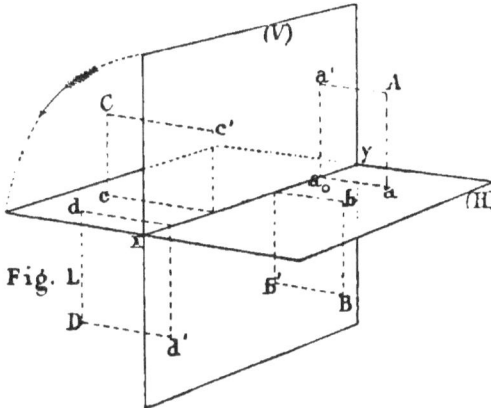

Fig. 1.

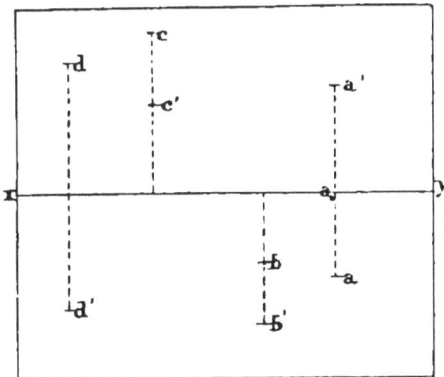

Fig. 2.

2—2

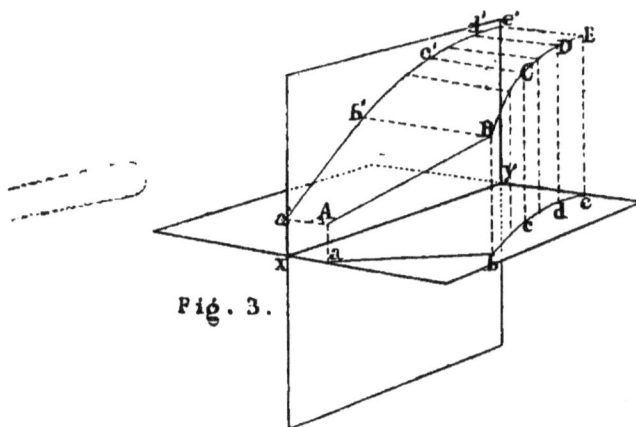

Fig. 3.

The point of intersection of a line with a surface or of one surface with another is called a *trace*. When not otherwise stated the term trace is used to denote the intersection of a line or surface with the planes of projection. The points A and B (figs. 9, 11, 13, 15) are the traces of the line AB: A is called the *horizontal trace*, and B the *vertical trace*. The straight lines LR and SN (fig. 5) are the vertical and horizontal traces of the plane LMN; bce (fig. 3) is the horizontal trace of the surface $BCEecb$, and $b'c'e'$ the vertical trace of the surface $BCEe'c'b'$.

The traces of a plane meet the ground line at the same point, which is the point of intersection of the ground line with the plane. Hence when the ground line is parallel to a plane it is parallel to the traces of that plane.

If a plane be parallel to one of the co-ordinate planes it cannot have any trace on that plane, and its trace on the other must be parallel to the ground line, as RS in fig. 7, (Theor. XII.).

When a plane is perpendicular to one of the co-ordinate planes its trace on the other is perpendicular to the ground line, Theor. VI. Thus LMN, fig. 7, is perpendicular to the horizontal plane and LM consequently perpendicular to xy.

Any two straight lines on the planes of projection which meet the ground line at the same point, or which are parallel to it, may be considered as the traces of a plane.

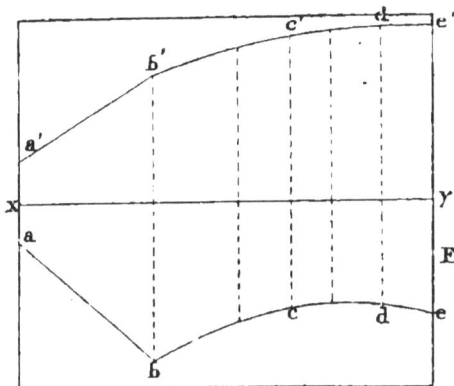

Fig. 4.

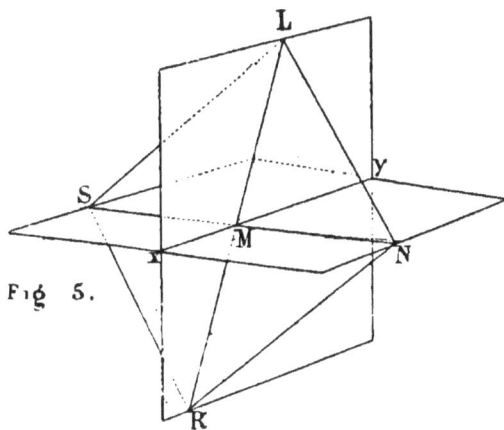

Fig. 5.

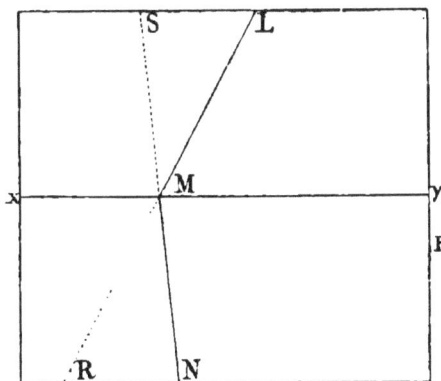

Fig. 6.

If a line be perpendicular to one of the co-ordinate planes its projection on the other will be perpendicular to the ground line. For its projecting plane will be perpendicular to the same plane to which the line is perpendicular (Theor. IV.), and therefore the trace of that projecting plane with the other co-ordinate plane must be perpendicular to the ground line.

Any two lines whatever, one on each plane of projection, may be the projections of the same line, except when they are both at right angles to the ground line, and meet it at different points; for in that case the two projecting planes being each perpendicular to the planes of projection (Theor. IV.) would also be perpendicular to the ground line (Theor. VI.), and therefore parallel to one another, Theor. XIII.

So far, points, lines, and surfaces, have been referred to two planes, but in order to work the problems it is necessary that these should be reduced to one. This is done by turning the vertical plane about the ground line as an axis till it coincides with the horizontal plane, the motion being in the direction of the arrow in fig. 1. This process is called the "rabatment"[1] of the vertical plane, which is supposed to carry with it all the points and lines which were conceived to be on it in the vertical position. As the ground line remains fixed, points and lines on the planes of projection will be respectively at the same distance from it and make the same angles with it after the rabatment of the vertical plane as before. Figs. 1, 3, 5, &c., show the planes of projection in their true positions; figs. 2, 4, 6, &c., show the points and lines on them after the rabatment of the vertical plane.

It has been shown that the perpendiculars from the projections aa' of a point A meet at the same point of the ground line, so that when the two planes coincide these two perpendiculars will be in one straight line, therefore

The projections of a point must lie in the same perpendicular to the ground line.

It will be seen from figs. 1 and 2 that a point is in front of the vertical plane as A, or behind it, as C, according as

---

[1] A term introduced by Dr Woolley as an equivalent for the French word Rabattement, which is used in the same sense by French writers on this subject.

its horizontal projection after the rabatment is below or above the ground line. Also a point is above or below the horizontal plane according as its vertical projection is above or below the ground line. It is evident too that *the distance of a point from the vertical plane is equal to the distance of its horizontal projection from the ground line, and its distance from the horizontal plane equal to the distance of its vertical projection from the ground line.*

It must be borne in mind that the rabatment of the vertical plane is only used as a means of practically working out the problem on a single plane, and in reasoning about, or in realizing to the mind, the relative positions of the points,

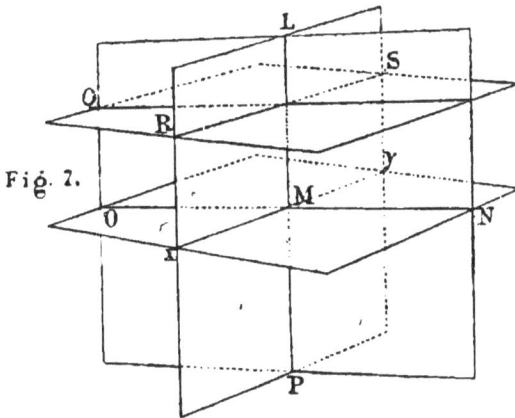

Fig. 7.

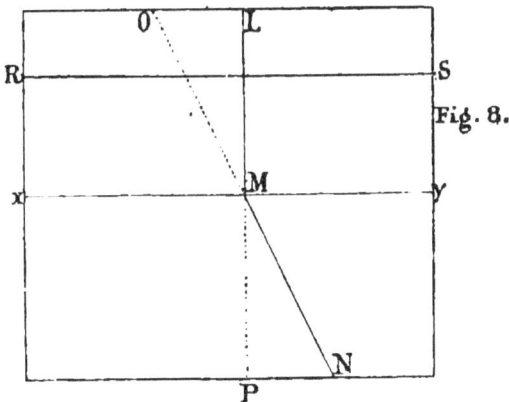

Fig. 8.

lines or figures in space, it is necessary to conceive the planes of projection at right angles to one another. To assist the student in forming a right conception of the following problems of this chapter two figures are given with each, one representing the principal points, lines and planes in their true positions with the co-ordinate planes at right angles to one another, and the other showing the actual solution of the problem on one plane.

PROBLEM I. Figs. 9 to 16.

Given the traces of a straight line, to find its projections.

Let A and B be the horizontal and vertical traces respectively of the straight line AB; it is required to find its projections.

Construction. From B draw Bb perpendicular to the ground line xy; the line Ab is the horizontal projection of AB.

Similarly, to find the vertical projection of AB, draw Aa' perpendicular to xy and join Ba'.

Proof. Since B is a point in the vertical plane of projection, the line Bb at right angles to xy is perpendicular to the horizontal plane (Theorem V.), and is therefore the horizontal projection of the point B of the given line. The horizontal projection of AB must pass through the point b, and it is evident it must pass through A; therefore Ab is the horizontal projection of AB.

It may be shown in a similar way that Ba' is the vertical projection of AB.

PROBLEM II. Figs. 9 to 16.

Given the projections of a straight line, to find its traces.

Let Ab, Ba' be the projections of a straight line AB; it is required to find its traces.

Construction. Produce the line Ab, if necessary, to meet the ground line at b; through the point b draw bB at right angles to xy to meet the vertical projection at B. The point B is the vertical trace of the given line.

To find the horizontal trace A, produce Ba' to meet the

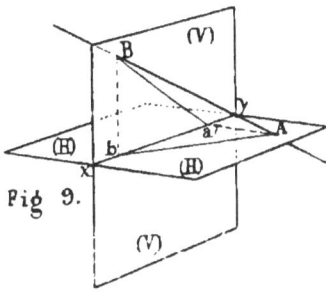

Fig 9.

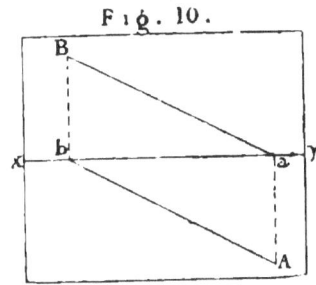

Fig. 10.

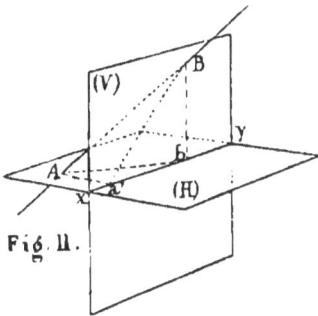

Fig. 11.

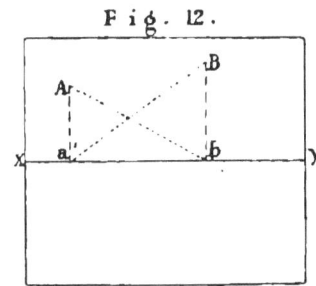

Fig. 12.

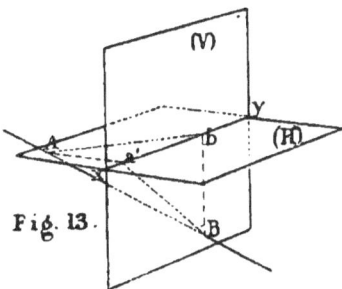

Fig. 13.

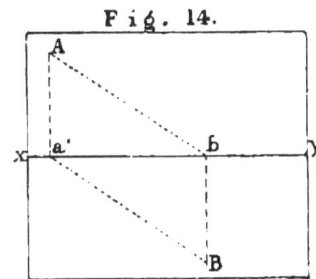

Fig. 14.

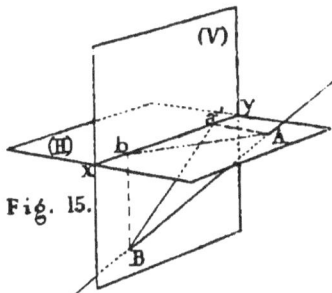

Fig. 15.

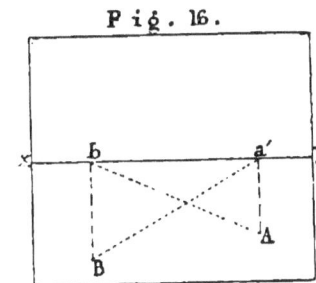

Fig. 16.

ground line, and draw $a'A$ perpendicular to xy to meet Ab in A. A is the horizontal trace of the line.

Proof. Because Bb is in the vertical plane of projection and is perpendicular to xy, it is perpendicular to the horizontal plane (Theor. v.); and since it passes through the point b, it must be the *projector* of the point in which the given line meets the vertical plane; that is, the vertical trace of the line is in Bb; but the vertical trace of the line must be in its vertical projection; it is therefore the point B in which the lines intersect.

Similarly it may be proved that A is the horizontal trace of the given line.

Corollary. If a straight line be parallel to one of the co-ordinate planes, its projection on the other will be parallel to the ground line.

PROBLEM III. Figs. 17 and 18.

To determine the projections of a straight line which passes through a given point, and is parallel to a given straight line.

Let Ab, $a'B$ be the projections of the given line AB, and pp' the projections of the given point P; it is required to draw the projections of the line which passes through P, and is parallel to AB.

Construction. Through the point p draw pC parallel to Ab, and through p' draw $p'c'$ parallel to $a'B$; pC and $p'c'$ are the projections required.

Proof. The line through P parallel to AB must have its horizontal and vertical projections respectively parallel to Ab and $a'B$ (Theor. XVI.), and passing through p and p', the projections of P; therefore pC and $p'c'$ are the projections required.

Corollary. As there can only be one line of which pC and $p'c'$ are the projections, it follows that if the vertical and horizontal projections of two straight lines are respectively parallel to one another, the lines are also parallel.

Fig. 17.

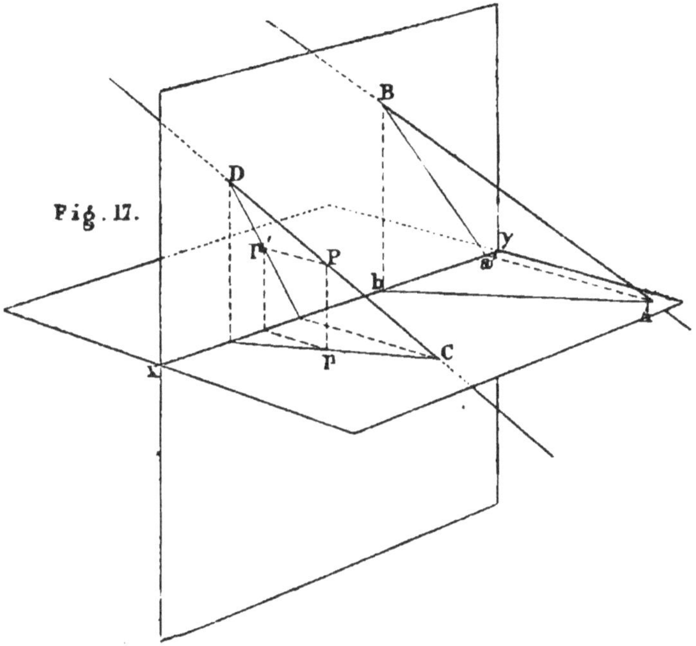

Fig. 18.

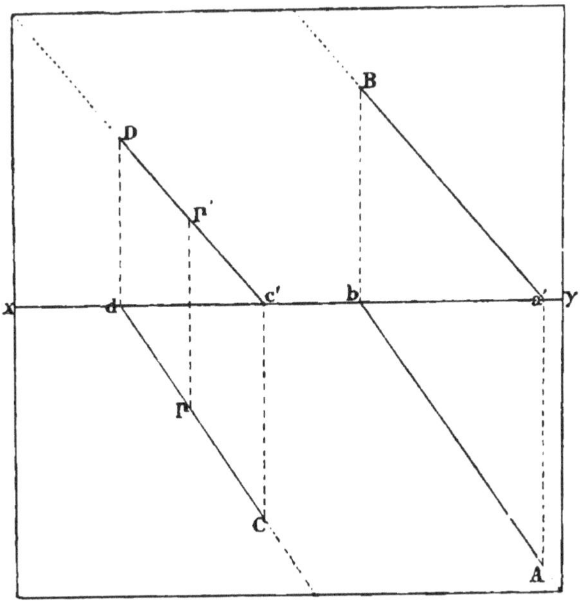

PROBLEM IV. Figs. 19 and 20.

Given the projections of two points, to determine the distance between them.

Let aa', bb' be the projections of two points A and B; it is required to find the length of the line AB.

It is obvious from figure 19 that AB is the hypotenuse of a right-angled triangle ABC of which AC is equal to ab, the horizontal projection of the line, and BC is the difference of the heights of A and B.   Hence the following :—

Construction. Through a' draw a line parallel to xy, meeting bb in the point c', and make $c'a_1'$ equal to ba : $b'a_1'$ is the distance required.

Proof. The right-angled triangle $a_1'b'c'$ has the two sides $a_1'c'$ and $b'c'$ equal respectively to AC and BC of the triangle ABC; therefore $a_1'b'$ is equal to AB.

Remarks. The triangle $a_1'b'c'$ may also be considered as the vertical projection of ABC when the vertical projecting plane of AB is turned about Bb, as an axis, till it comes parallel to the vertical plane of projection, when every line of the figure $AabB$ is equal to its vertical projection (Theor. xv. Cor. 1).

It may be observed that the problem might be solved in a similar manner by constructing a right-angled triangle with its base equal to $a'b'$, and the perpendicular equal to the *difference* of Aa' and Bb'.   If A and B were on opposite sides of the vertical plane, the perpendicular would be the *sum* of the projectors.

PROBLEM V. Figs. 19, 20, 21, 22.

Given the projections of a point, and a line through the point, to lay off a given distance from the point along the line.

Let bh, $b'h'$ be the projections of the line, and bb' the projections of the point ; it is required to find the projections of a point on BH at a given distance from B.

Construction. Take any point H on the line and con-

Fig. 19.

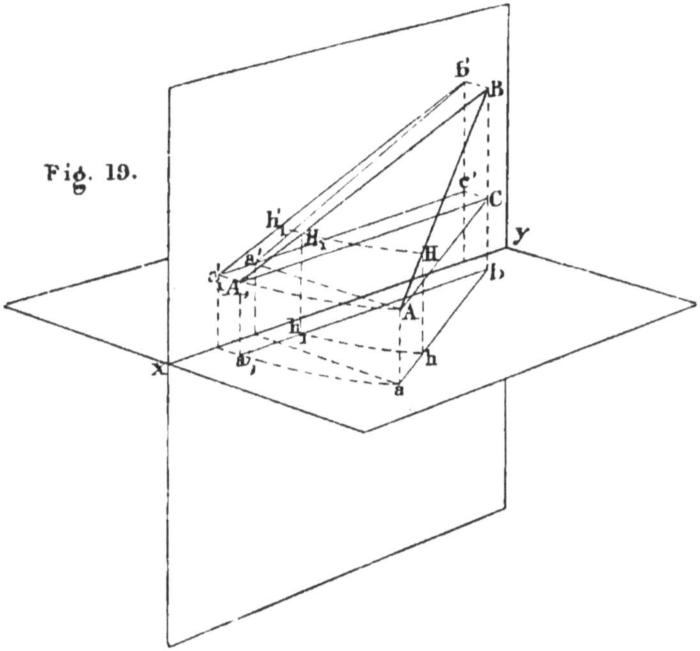

Fig. 20.

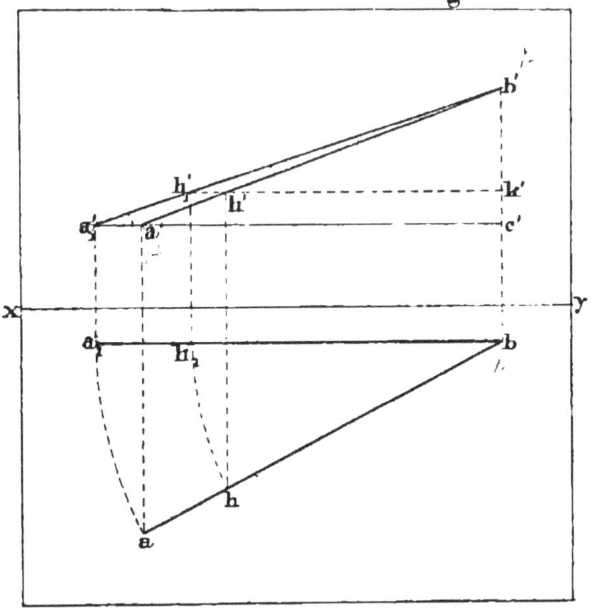

struct the triangle $b'h_1'k'$ as in Problem IV. ; set off the given distance $b'a_1'$ from b' along the line $b'h_1'$, produced if necessary; draw $a_1'a_1$ perpendicular to xy to meet ba_1 drawn parallel to xy, and make ba equal to ba_1. a will be the horizontal projection of the point required, and a' its vertical projection.

$$Proof. \qquad \frac{b'h_1'}{b'a_1'} = \frac{bh_1}{ba_1},$$

since bb', h_1h_1' and a_1a_1' (figs. 20 and 22) are parallel to one another.

That is
$$\frac{b'h_1'}{b'a_1'} = \frac{bh}{ba}.$$

(In figs. 21 and 22, H and h coincide.)

Also
$$\frac{BH}{BA} = \frac{bh}{ba} \dots\dots \text{Theorem XV, Cor. 2.}$$

Therefore
$$\frac{BH}{BA} = \frac{b'h_1'}{b'a_1'}.$$

But $BH = b'h_1'$; therefore $BA = b'a_1'$, the given length.

Remarks. It will be seen from this proof that ba is a fourth proportional to the three lines BH, bh, and BA, the given length.

Another way of considering the problem, which gives the same construction, is to suppose Bbh, the vertical projecting plane of BH, to turn about Bb as an axis till it is parallel to the vertical plane of projection, and finding the vertical projection bh_1' of the given line in that position; then setting off the given distance from b' along $b'h_1'$ which gives a_1 as the horizontal projection of the point required when the line BA is parallel to the vertical plane. The plane Bbh is then returned to its former position. In the motion of Bbh about Bb any line ba or CA at right angles to Bb will move in a horizontal plane (Theor. III. Cor. 2).

This manner of considering the solution will be easily understood from the figures.

Fig. 21.

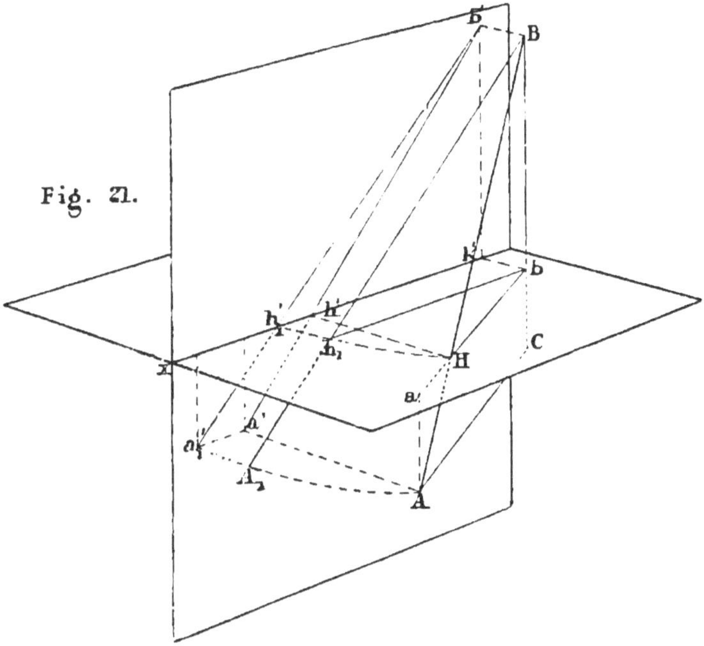

Fig. 22.

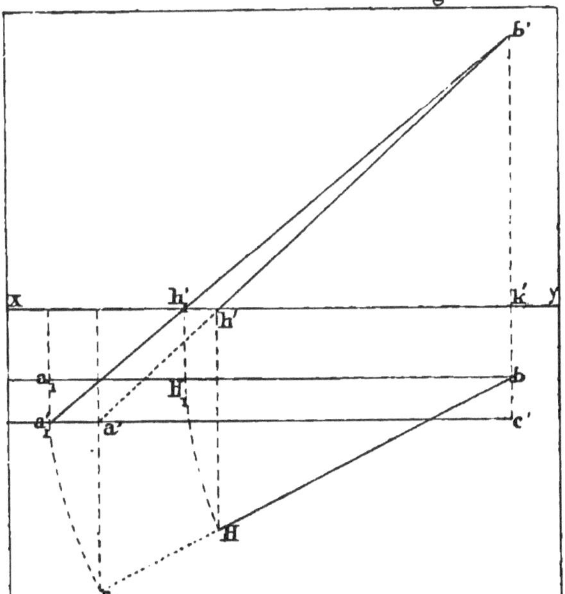

PROBLEM VI.   Figs. 19, 20, 21, 22.

Given the projections of a line, to find the angles which it makes with the planes of projection.

Let it be required to find the angle which AB makes, (1) with the horizontal plane, (2) with the vertical plane.

Construction.   Draw the triangle $b'a_1'c'$ as in Problem IV; the angle $b'a_1'c'$ will be the angle which AB makes with the horizontal plane.

Proof.   Because the triangle $b'a_1'c'$ is constructed equal to BAC, the angle $b'a_1'c' = BAC$.

But since CA is parallel to ab, BAC is equal to the angle between AB and ab.

Therefore $b'a_1'c'$ is the angle required.

If a right-angled triangle be constructed having $a'b'$ for one side, and the difference of Aa' and Bb' for the other, the angle between the hypotenuse and $a'b'$ will be the angle which AB makes with the vertical plane of projection.

When the line meets a plane of projection as in figs. 21 and 22, it is generally most convenient to construct the triangle $b'h_1'k'$ for determining the angle which it makes with that plane.

Problem VII. Figs. 23, 24.

To determine the projections of a straight line which shall contain a given point and make given angles with the planes of projection.

Let pp' be the projections of a point P; it is required to find the projections of a straight line passing through P, making an angle α with the horizontal plane, and an angle β with the vertical plane of projection.

Construction. Draw $p'a'$ making an angle α with xy; with centre p and radius equal to oa' describe the circle AA_1. Draw the line $p'b$ making the angle $a'p'b$ equal to β, and from a' draw $a'b$ perpendicular to $p'b$; with centre p' and radius $p'b$ describe a circle meeting xy in a_1', and draw $a_1'A_1$ perpendicular to xy to meet the circle AA_1 in the point A_1. pA_1 and $p'a_1'$ will be the projections of the line required.

Proof. Since in the two right-angled triangles PpA_1 and $p'oa'$, $Pp = p'o$ and $pA_1 = oa'$, the triangles are equal in every respect.

Therefore the angle $PA_1p = p'a'o = a$.

Again, the line PQ is parallel to $p'a_1'$ (Prob. III. Cor.), and therefore equal to it (Theorem XV. Cor.), and consequently equal to $p'b$.

Then in the two right-angled triangles PQA_1 and $p'ba'$ $PA_1 = p'a'$ and $PQ = p'b$.

Therefore the angle $A_1PQ = a'p'b = \beta$. ´

But the angle A_1BQ is equal to the angle between PA_1 and $p'a_1'$, since PQ and $p'a_1'$ are parallel.

Therefore the inclination of PA_1 to the vertical plane is equal to β.

Remarks. It may be proved in a similar manner, that the three lines passing through P and the points marked 1, 2, 3 make the given angles with the planes of projection; so that, in general, there are four solutions to this problem. If β were the complement of α, and therefore equal to $op'a'$, $p'b$ would be equal to $p'o$ and the circle described from the

Fig 23.

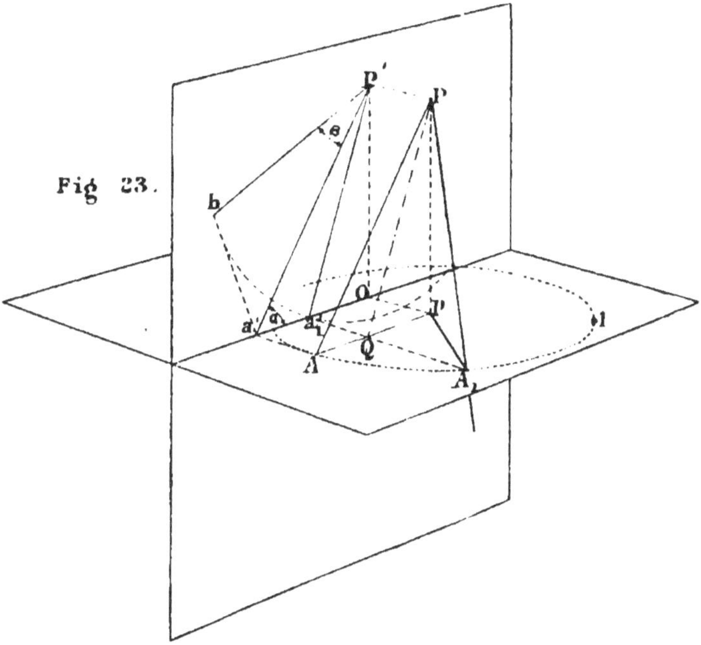

Fig. 24.

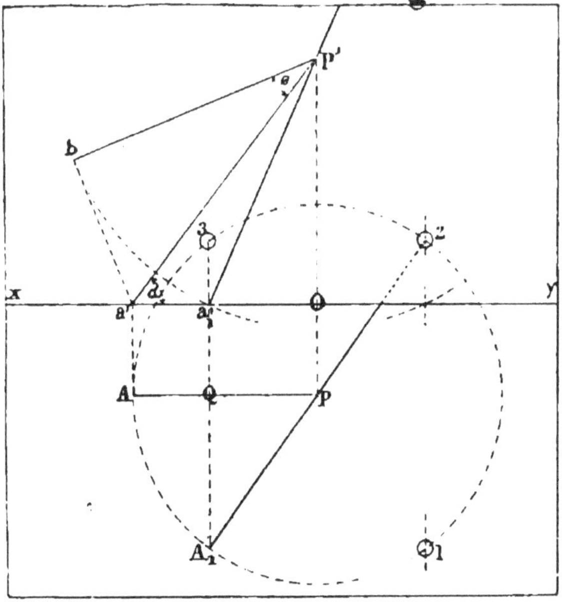

centre p' through b would meet xy in one point only; in that case there would be two solutions.

If $\alpha + \beta$ were greater than 90°, $p'b$ would be less than $p'o$ and the circle described through b would not meet xy, and the solution would be impossible. That is, the sum of the angles which a line makes with the planes of projection cannot exceed 90°; and when they are equal to 90° the projections of the line are at right angles to the ground line.

PROBLEM VIII. Figs. 25 and 26.

Given the projections of two straight lines which intersect, to find the angle between them.

Let ab, $a'b'$ and ac, $a'c'$ be the projections of the two straight lines AB, AC; it is required to find the angle BAC.

FIRST SOLUTION. *Construction.* Draw any line $b'c'$ parallel to xy, and draw $b'b$, $c'c$ perpendicular to xy to meet the horizontal projections of the two given lines. Join bc. Find the lengths of the lines AB and AC (Prob. IV.), and on bc, as base, describe the triangle $ba_{1}c$, having the sides ba_{1}, ca_{1} equal to BA and CA respectively. The angle $ba_{1}c$ will be the angle required.

Proof. The line $bc = BC$, since BC is horizontal (Theor. XV. Cor. 1). Therefore the three sides of the triangle $ba_{1}c$ are respectively equal to the three sides of the triangle BAC. Hence the angle $ba_{1}c = BAC$.

SECOND SOLUTION. *Construction.* Find the projections of the horizontal line BC as before. Draw ad perpendicular to bc, and on da produced set off da_{1} equal to the line DA. The angle $ba_{1}c$ will be the angle required.

Proof. ad and $a'd'$ are the projections of the perpendicular from A on the base BC (Theorem XVII.).

Since in the two triangles BAC, $ba_{1}c$,

$$BD = bd, \; DC = dc \text{ and } DA = da_{1},$$

the triangles are equal in every respect, and consequently the angle $ba_{1}c = BAC$.

Fig. 25.

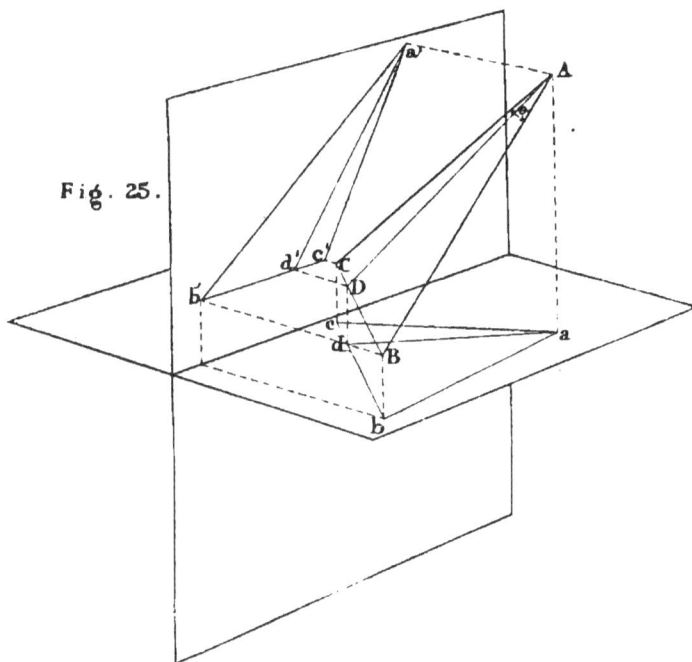

Fig. 26.

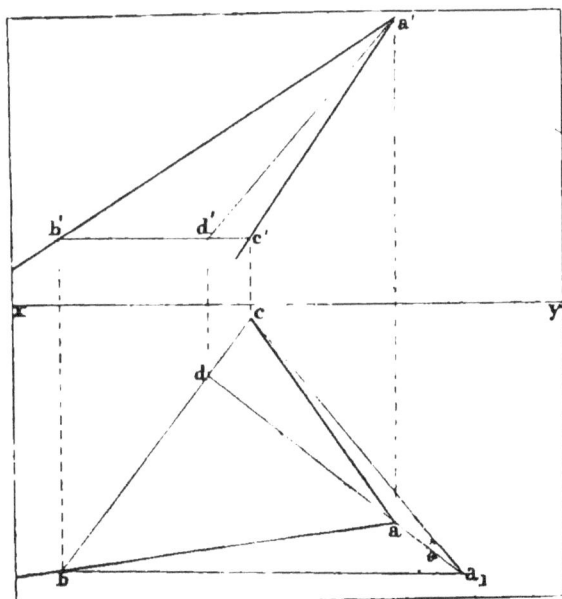

If the two given lines were AB and BC, either AC or AD could be drawn so as to form a triangle from which the angle ABC might be determined.

Corollary. *To bisect the angle between two given lines.*

Construct the angle θ between the lines, and find the point where the bisector meets the base BC; then join the projections of that point with a and a' respectively.

PROBLEM IX.   Figs. 27 and 28.

To determine the horizontal projection of a given angle θ, when the lines containing it make angles of α and β respectively with the horizontal plane.

Construction. From any point A in the vertical plane of projection draw AB and AC, making angles α and β respectively with xy. Also draw AD equal to AC and making an angle θ with AB. Construct the triangle BaC_1, having the sides aC_1, C_1B equal respectively to aC and BD. BaC_1 will be the horizontal projection required.

Proof. It is obvious from fig. 27 that the triangle AaC_1 is equal to AaC, and therefore AC_1 has the required inclination, β.

Also since $BC_1 = BD$, and $AC_1 = AD$, the triangle $ABC_1 = ABD$, and consequently $BAC_1 = \theta$.

Remarks. This is called reducing an angle to the horizon. It may be used in mapping for finding the horizontal projections of angles measured with the sextant between objects at different altitudes.

If the projection of one of the lines be given in position, the vertical plane in which AB and AC are drawn should contain that line.

$\alpha + \beta + \theta$ cannot exceed $180°$.

Fig. 27.

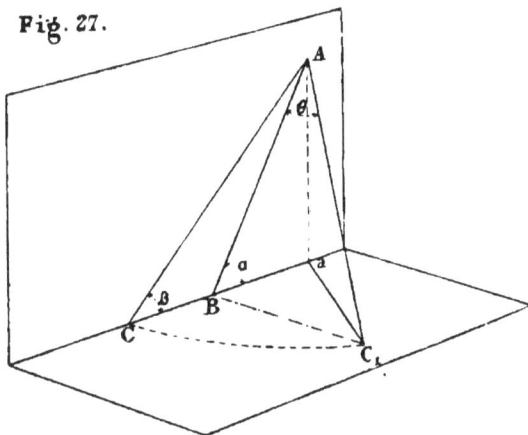

Fig. 28.

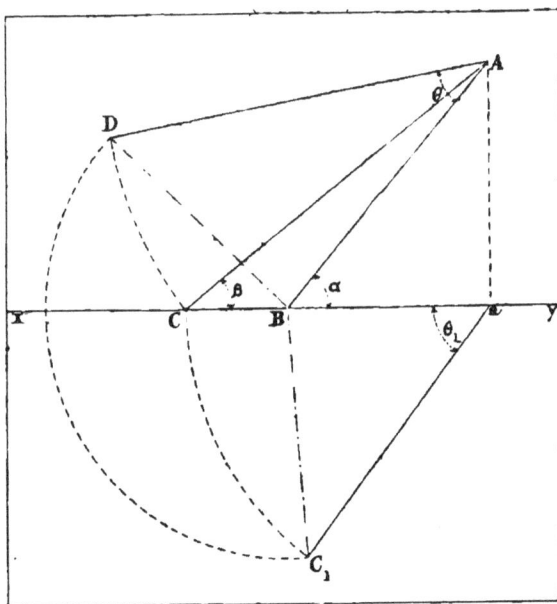

PROBLEM X.   Figs. 29 and 30.

To determine the traces of a plane containing three given points not in the same straight line.

Let *aa′*, *bb′*, *cc′* be the projections of three points *A, B, C*; it is required to find the traces of the plane containing *A, B* and *C.*

Construction.   Find the traces *L* and *M* of the line *AB* (Prob. II.) and the traces *Q* and *R* of *BC*.   The lines *LQ* and *MR* will be the traces of the plane required.

Proof.   Since the points *L* and *M* are in the plane *LSM*, the whole line *LM* is in it (def. 3), and therefore the points *A* and *B*.

Similarly since *R* and *Q* are in the plane *LSM*, *B* and *C* must be in it.

Therefore the plane *LSM* contains *A, B* and *C*.

Remarks.   The traces of the third line *AC* must also be in the traces of the plane.

As the traces of a plane meet on the ground line, it follows that one trace of a plane and a point on the other trace is sufficient to determine it.

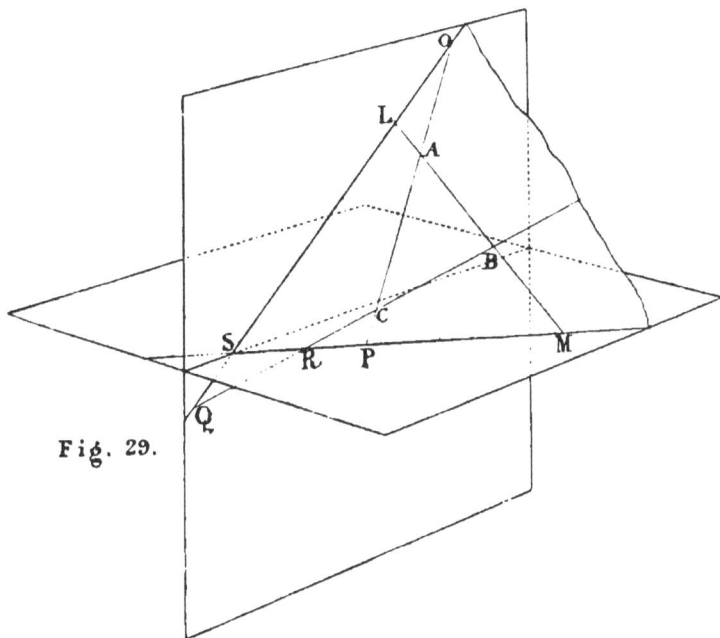

Fig. 29.

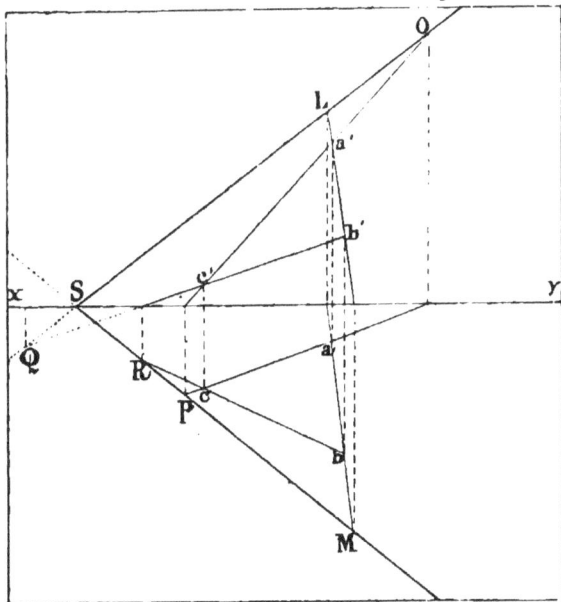

Fig. 30.

PROBLEM XI.   Figs. 31 and 32.

Given the projections of two lines which are not in the same plane, to determine the traces of a plane which shall contain one of these lines and be parallel to the other.

Let AB and PQ be the given lines, it is required to find the traces of a plane containing AB and parallel to PQ.

Construction. Find the traces L and M of the line AB. Through any point of AB, as B, draw a line parallel to PQ (Prob. III.) and determine N and O the traces of this line. $LNSO$ will be the plane required.

Proof. Because the plane LSO contains NO, which is parallel to PQ, therefore the plane is parallel to PQ (Theor. X.); and as LSO contains AB, it is the plane required.

Fig. 31.

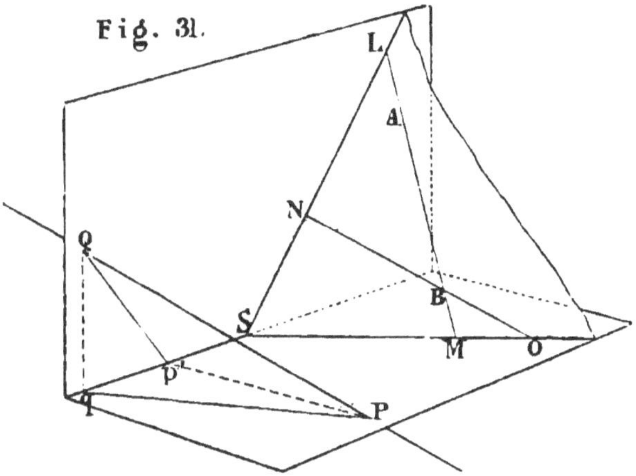

Fig. 32.

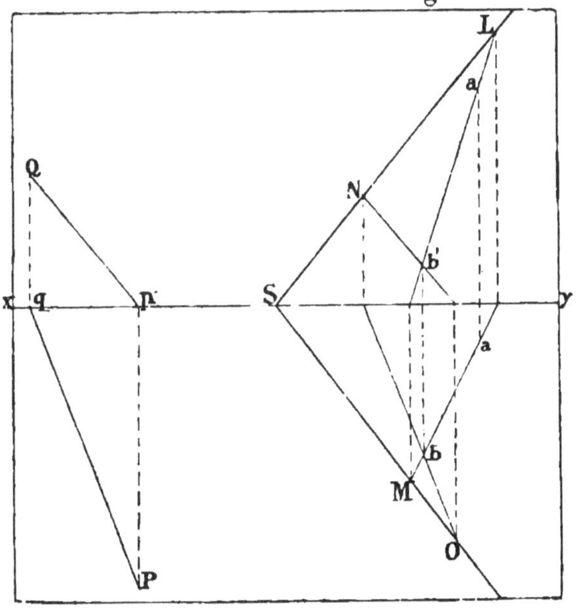

PROBLEM XII. Figs. 33 and 34.

To determine a plane which shall contain a given point and be parallel to a given plane.

Let it be required to find the traces of a plane containing the point *P* and parallel to *LMN*.

Construction. Through *P* draw *PQ* parallel to *MN*, and find its vertical trace *Q*. Draw *QS* and *ST* parallel respectively to *LM* and *MN*. The plane *RST* will be the one required.

Proof. Because *RS* and *ST* are parallel respectively to *LM* and *MN* the planes are parallel (Theor. XIV.).

Again, since *PQ* and *ST* are both parallel to *MN* they are parallel to one another (Theor. IX.), and therefore in the same plane; so that the plane *RST* which contains *Q* and *ST* must also contain *P*.

Remark. The line through *P* need not necessarily be drawn parallel to *MN*, but may be parallel to any line whatever in the plane *LMN*.

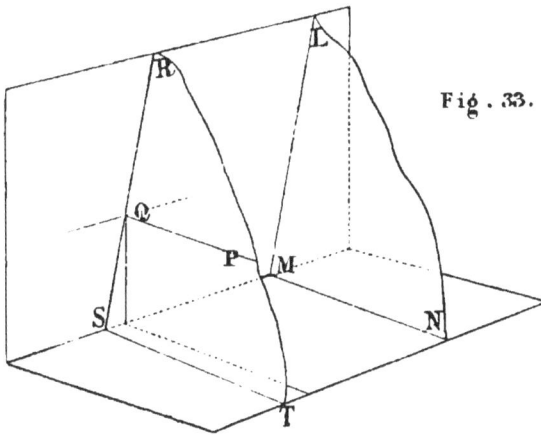

Fig . 33.

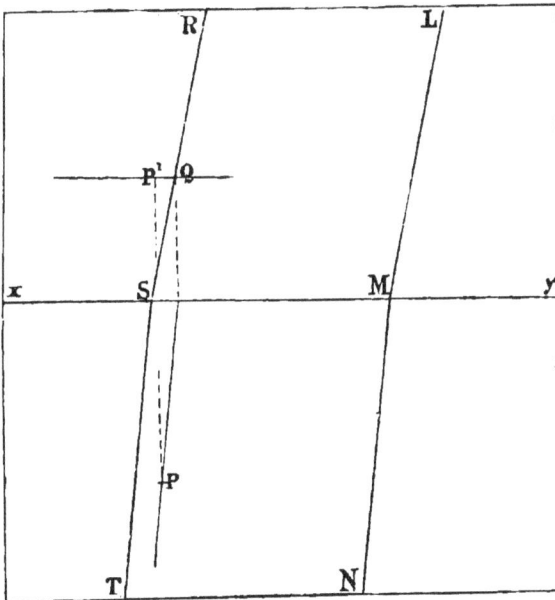

Fig. 34.

Given the traces of two intersecting planes, to draw the projections of their common section.

Let it be required to find the projections of the common section of the planes LMN and OPQ.

Construction. Find A and B, the points of intersection of the vertical and horizontal traces respectively of the planes, and draw the projections of the line whose traces are A and B; these will be the projections required.

Proof. It is evident that A and B are points of the common section; and since they are points in the planes of projection they are the traces of the common section.

Corollary 1. If three planes have a common point it must lie in the line of intersection of each pair of planes; hence the projections of the point may be determined by finding the projections of any two of these lines.

Thus pp' are the projections of the point of intersection of the three planes marked (1), (2), (3) (figs. 37 and 38).

Cor. 2. When the two planes meet the ground line at the same point, as planes (1) and (2) (figs. 37 and 38), the method given above fails. A point in the common section of the two planes may be found by taking a third plane, (3), intersecting the other two and finding the common point P of the three planes. Then as the common section of (1) and (2) must also pass through M it is completely determined.

Cor. 3. If the traces on one plane of projection only be parallel then the common section of the two planes must be parallel to these (Theor. XI. Cor. 2). Hence it is only necessary to draw the projections of a line passing through the point of intersection of the traces on one plane of projection and parallel to those on the other.

Cor. 4. If both traces of each plane be parallel to the ground line the common section will also be parallel to the ground line (Theor. XI. Cor. 2, and Theor. IX.). Therefore one point of the line will be sufficient to determine it. This point may be found by taking a third plane intersecting the other two and finding the common point.

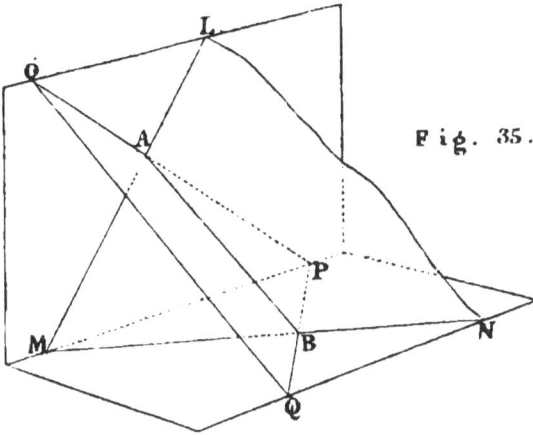

Fig. 35.

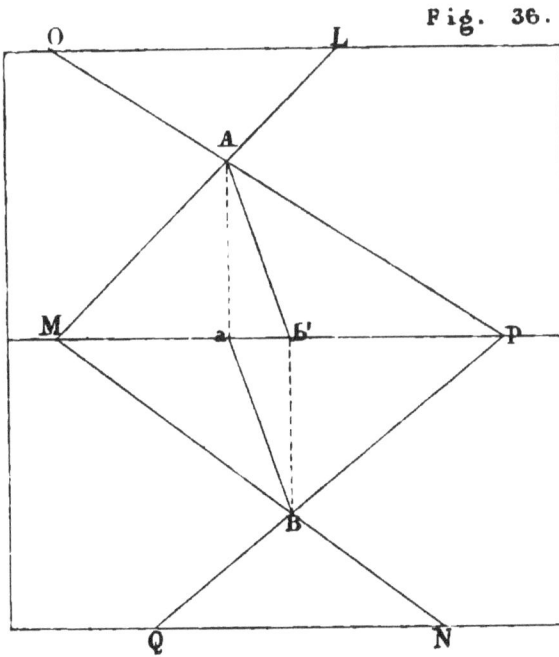

Fig. 36.

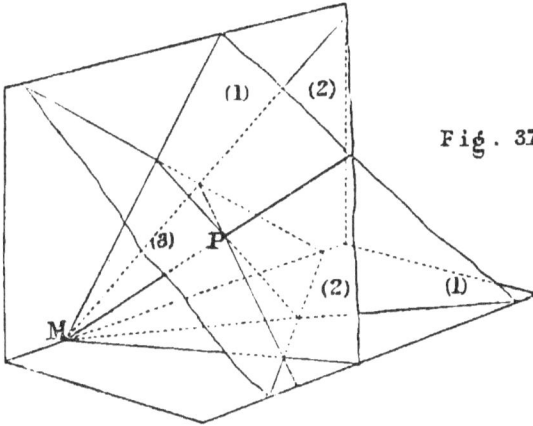

Fig. 37.

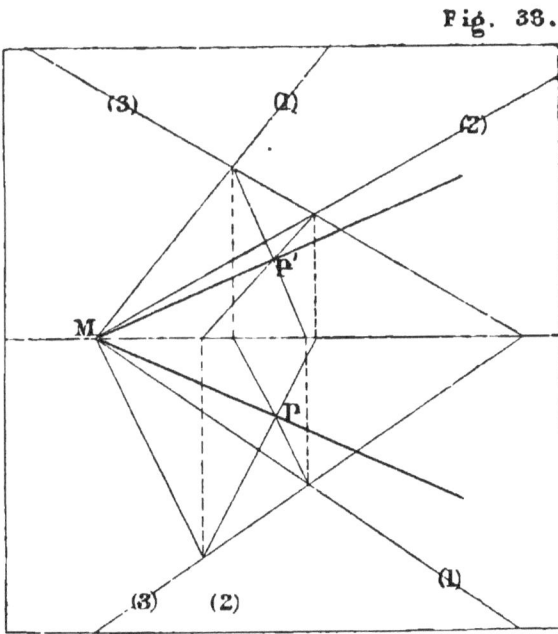

Fig. 38.

PROBLEM XIV. Figs. 39 and 40.

To find the point of intersection of a line and a plane.

Let ab, $a'B$ be the projections of the line and LMN the traces of the plane.

Construction. Take the vertical plane AaB containing the line AB, and find the vertical projection Cd' of the common section of AaB and LMN. The point o' where $a'B$ and Cd' intersect will be the vertical projection of the point required; the other projection will be o on ab.

Proof. Since AB and CD are in the same plane AaB they intersect in the point O. But CD is also in the plane LMN, and therefore O is in that plane.

Hence AB meets LMN in O.

Remark. Any plane whatever might be taken containing AB; but when the plane is taken perpendicular to one of the planes of projection, as above, the solution is shortest.

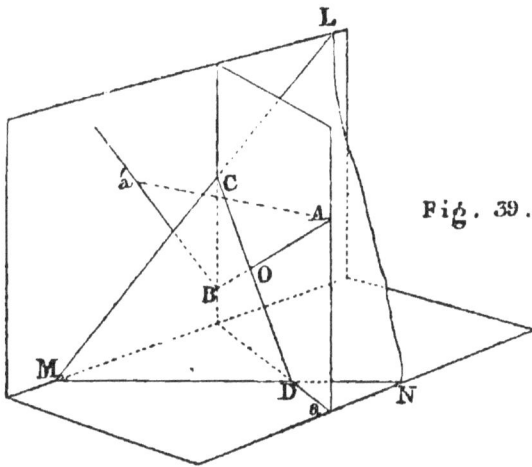

Fig. 39.

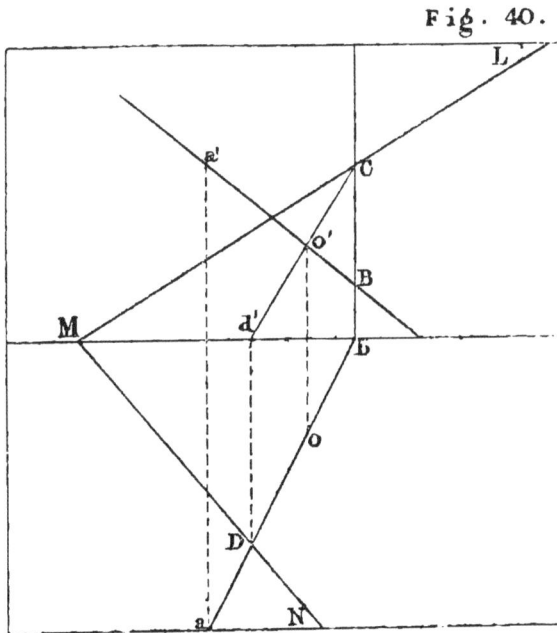

Fig. 40.

To determine the projections of a line which shall contain a given point and be perpendicular to a given plane.

Let pp' be the projections of the point and LMN the traces of the plane.

Construction. Draw pq perpendicular to MN and $p'q'$ perpendicular to LM; pq, $p'q'$ will be the projections of the line containing P and perpendicular to LMN.

Proof. Because MN is horizontal it is perpendicular to Pp, and since it is also perpendicular to pS it is perpendicular to the plane PQS (Theor. III. Cor. 1), and therefore the plane LMN is perpendicular to PQS (Theor. IV.).

It may be proved in a similar manner that LMN is perpendicular to the projecting plane $p'PQ$.

Therefore, since PQS and $p'PQ$ are perpendicular to LMN, their common section, PQ, is perpendicular to it (Theor. VI.).

Cor. 1. *To determine the distance from a point to a plane.*

Draw the projections of the perpendicular and find the point of intersection Q by Problem XIV.; then determine the length of the line PQ (Prob. IV.).

Cor. 2. *To determine a plane parallel to a given plane and at a given distance from it.*

From a point in the given plane draw a line at right angles to it and set off the given distance from that point along the line (Prob. V.); then through the point so found draw a plane parallel to the given plane (Prob. XII.).

Cor. 3. *To determine the traces of a plane which shall contain a given line and be perpendicular to a given plane.*

From any point of the given line draw a perpendicular to the given plane, and find the traces of the plane containing these two intersecting lines (Prob. X.). This will be the plane required (Theor. IV.).

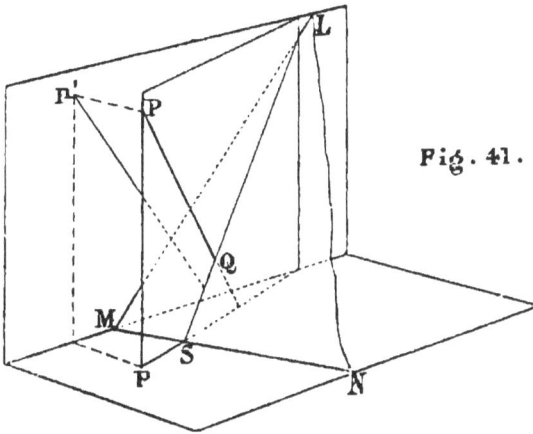

Fig. 41.

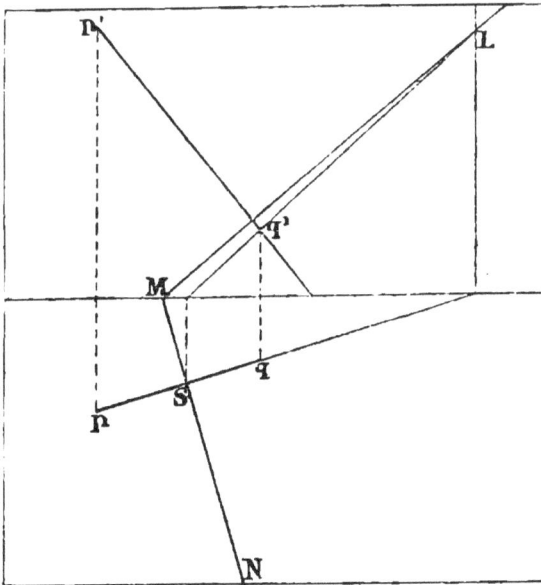

Fig. 42.

Cor. 4. *To determine the traces of a plane which shall contain a given point and be perpendicular to two given planes.*

From the given point draw a perpendicular to each of the given planes, and find the traces of the plane containing these two perpendiculars (Prob. x.).

Note. It has been proved in this problem that when the projections of a line are at right angles to the traces of a plane the line is perpendicular to the plane. The converse is also true; *i.e.* if a line be perpendicular to a plane the projections of the line are at right angles to the traces of the plane. For, otherwise two perpendiculars might be drawn to a plane from the same point, which is impossible (Theor. III. Cor. 3).

PROBLEM XVI.   Figs. 43 and 44.

To determine the traces of a plane which shall contain a given point, and be perpendicular to a given line.

Let pp' be the projections of a point P, and ab, $a'b'$ the projections of a line AB; it is required to find the traces of a plane containing P and perpendicular to AB.

It follows from Problem XV., that the traces of the required plane must be at right angles to ab, $a'b'$; so that it is only necessary to determine a point in one of the traces: hence the following :—

Construction. Through P draw a line parallel to the horizontal trace of the required plane, that is a line PQ having its horizontal projection pq at right angles to ab, and its vertical projection parallel to the ground line, and find the vertical trace Q of this line. Through Q draw QM perpendicular to $a'b'$, and through M draw MN perpendicular to ab. QMN will be the plane required.

Proof. Because the lines PQ and MN are parallel they are in the same plane; but the point Q is in the plane QMN, therefore the whole line PQ is in that plane.

That is the plane QMN contains P, and it is perpendicular to AB (Prob. XV.).

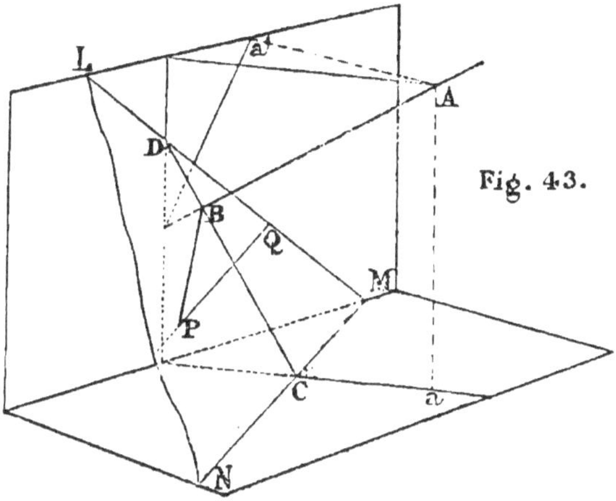

Fig. 43.

Fig. 44.

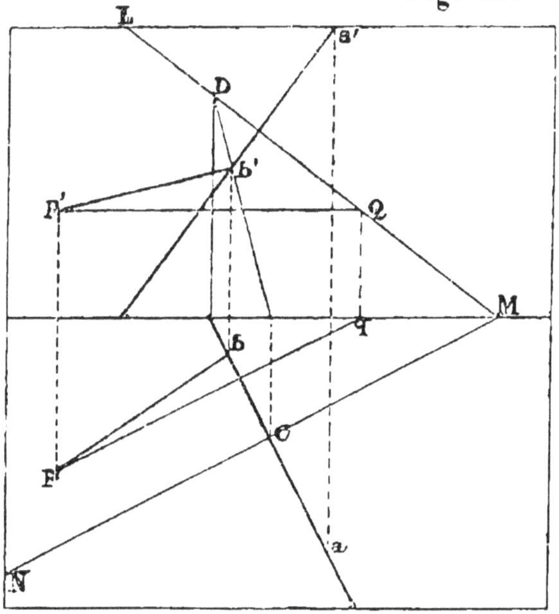

PROBLEM XVII.   Figs. 43 and 44.

From a given point to draw a perpendicular to a given line.

Let pp' be the projections of the point, and ab, $a'b'$ the projections of the line; it is required to find the projections of the perpendicular from P on AB.

Construction.   Draw a plane LMN containing P, and perpendicular to AB (Prob. XVI.), and determine the point of intersection B (Prob. XIV.). PB will be the perpendicular required.

Proof.   Because AB is perpendicular to the plane LMN, it is also perpendicular to PB (Def. 5).

PROBLEM XVIII.   Figs. 45 and 46.

To determine the common perpendicular to two straight lines which are not in the same plane.

Let AB and CD be the two given lines; it is required to find their common perpendicular.

Construction.   Find the traces of the plane DMC containing CD and parallel to AB (Problem XI.).

Find the traces of the plane ANB containing AB and perpendicular to the plane DMC (Prob. XV. Cor. 3).

Next determine the projections of the common section EF of the two planes DMC and ANB; and from the point G, where EF intersects CD, draw GH perpendicular to the plane DMC (Prob. XV.). GH will be perpendicular to the two lines AB and CD.

Proof.   Because GH is perpendicular to the plane DMC it is perpendicular to CD (Def. 5).   For the same reason it is perpendicular to EF, and therefore lies in the plane ANB (Theor. V. Cor.).

But AB and EF are parallel (Theor. XI.), and since GH is perpendicular to EF it must be also perpendicular to AB.

Therefore GH is the common perpendicular to AB and CD.

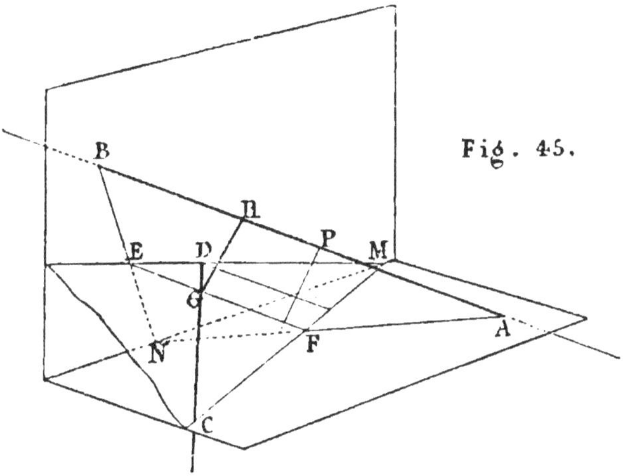

Fig. 45.

Fig. 46.

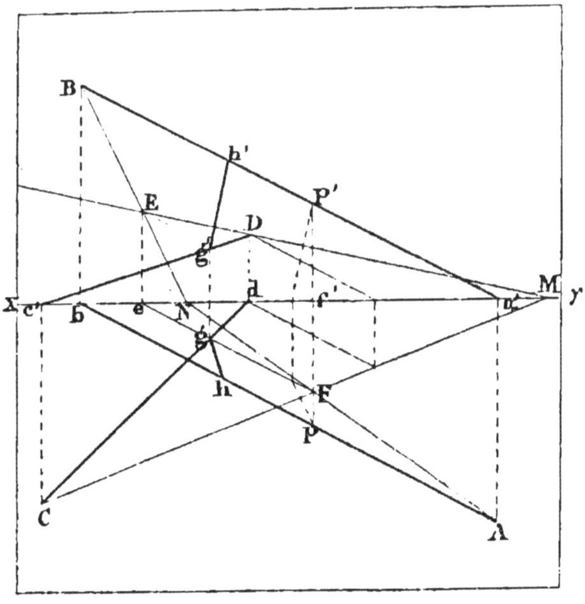

PROBLEM XIX. Figs. 47 and 48.

To determine the inclination of a plane to each plane of projection.

Let AMD be the traces of a plane, it is required to find its inclination (1) to the horizontal plane; (2) to the vertical plane.

Construction. Draw a vertical plane AaB, having its horizontal trace aB at right angles to MD, and find the angle between AB, the common section of the two planes, and aB (Prob. VI.); that will be the inclination of AMD to the horizontal plane.

To find the inclination to the vertical plane; draw the traces of a plane $Cq'D$, perpendicular to the vertical plane, and having its vertical trace Cq' at right angles to AM; the angle β, between DC and Cq', will be the inclination required.

Proof. Since AaB is perpendicular to the horizontal plane, and DB is perpendicular to aB, it is also perpendicular to the plane AaB (Theor. V.), and consequently AB is at right angles to BD (Def. 5). Because AB and aB, one in each plane, are both perpendicular to MD, the common section of the two planes, the angle ABa measures the dihedral angle between the planes (Def. 6).

In a similar manner it may be proved that the inclination of AMD to the vertical plane is equal to the angle between DC and Cq'.

PROBLEM XX. Figs. 47 and 48.

To find the traces of a plane which shall contain a given point, and make given angles with the planes of projection.

Let P be the given point, α and β the required inclinations to the horizontal and vertical planes respectively.

Construction. Through the point P draw a line PQ, making an angle of $(90^\circ - \alpha)$ with the horizontal and an angle of $(90^\circ - \beta)$ with the vertical plane of projection (Prob. VII.).

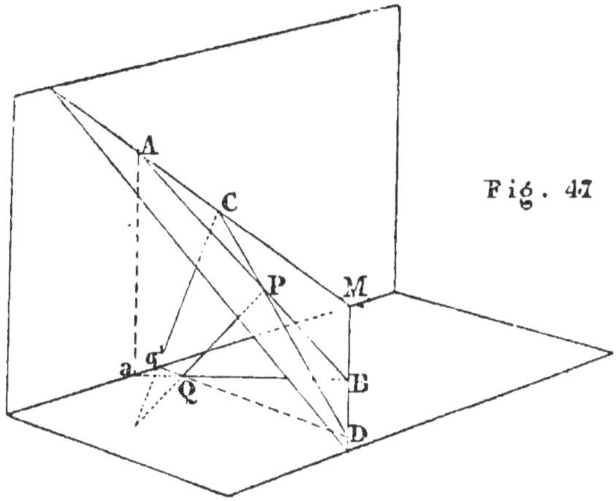

Fig. 47

Fig. 48.

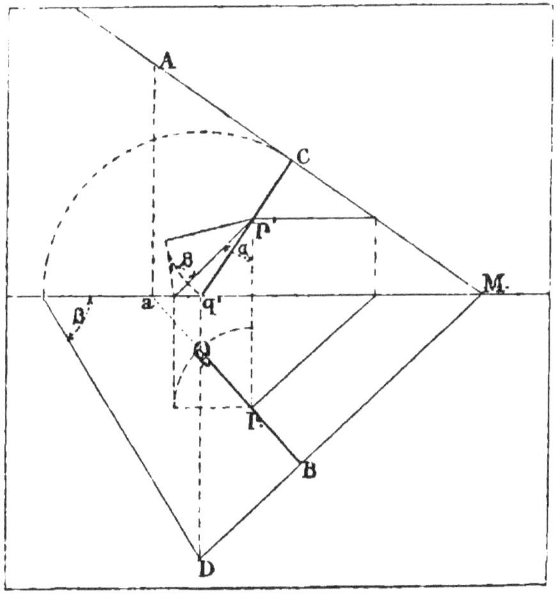

Next, determine the traces AM and MD of a plane containing P and perpendicular to PQ (Prob. XVI.). AMD will be the plane required.

Proof. Because PQ is perpendicular to AMB, the plane PBQ is also perpendicular to AMB (Theor. IV.), and since it is the projecting plane of PQ, it is perpendicular to the horizontal plane. Therefore the inclination of AMB to the horizontal plane is measured by the angle PBQ (Def. 6).

But since PQ is perpendicular to AMB, QPB is a right angle (Def. 5), and therefore PQB, PBQ are complementary, and since PQB has been constructed equal to $(90^\circ - \alpha)$, $PBQ = \alpha$.

In a similar way it may be proved that AMB is inclined at an angle β to the vertical plane of projection.

Remark. As four lines can be drawn, making the angles $(90^\circ - \alpha)$ and $(90'' - \beta)$ with the co-ordinate planes (Prob. VII.), so there may be four planes drawn to fulfil the conditions of this problem.

PROBLEM XXI. Figs. 49 and 50.

Given the traces of two intersecting planes to determine the angle between them.

Let LMN, LON be two planes intersecting in LN; it is required to determine the dihedral angle between them.

If a plane, as BAC fig. 49, were drawn perpendicular to LN, the angle BAC would be the angle required (Def. 5 and 6). Hence the following:—

Construction. Find lN, the horizontal projection of LN. Draw any line BC at right angles to lN and cutting it in D. Determine the perpendicular to LN from D: this may be done by constructing the triangle LlN_1 on the vertical plane of projection equal to LlN, and then drawing the perpendicular from D_1 on LN_1, as in fig. 50.

Set off along the line Dl the distance DA_2 equal to D_1A_1; and join A_2 with B and C. BA_2C is the angle required.

Proof. Because BD is perpendicular to lN, it is perpendicular to the plane LlN (Theor. v.) and therefore to the line LN (Theor. iii. Cor. 1). But DA is also perpendicular to LN; therefore LN being at right angles to BD and DA, is at right angles to the plane BAC; consequently the angle BAC is the angle required.

But the triangle BA_2C has been constructed equal to BAC.

Therefore the angle BA_2C is equal to the angle between the two planes.

PROBLEM XXII. Figs. 49 and 50.

To determine the traces of a plane which shall intersect a given plane in a given line, and make a given angle with it.

Let LON be the given plane and LN the given line in it; it is required to find the traces of a plane containing LN and making an angle θ with LON.

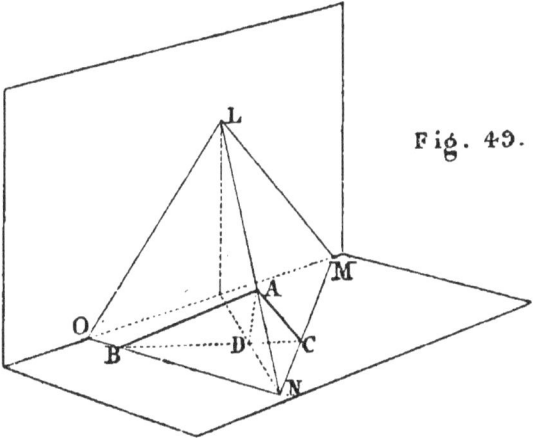

Fig. 49.

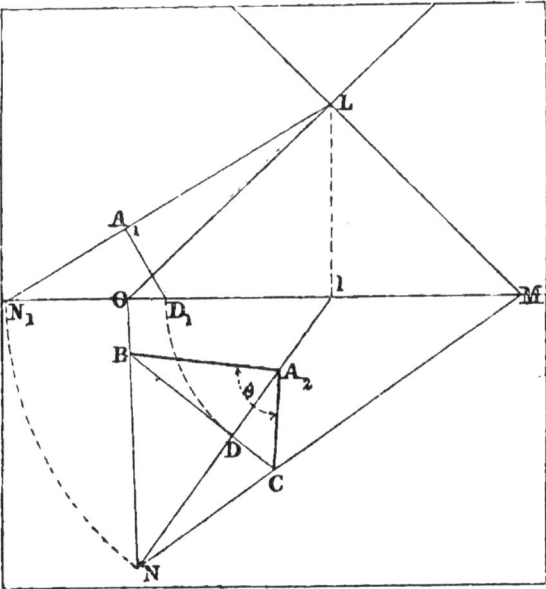

Fig. 50.

Construction. Draw any line BC perpendicular to LN, and determine the perpendicular DA as in Prob. XXI. Make DA_2 equal to DA; join BA_2, and make the angle BA_2C equal to θ. Draw NC, producing it to meet the ground line in M, and join ML. LMN will be the plane required.

Proof. It may be proved, as in Problem XXI., that the angle between the two planes LMN and LON is equal to BA_2C; but BA_2C is equal to the given angle θ.

Also LMN contains LN since it contains the points L and N. Therefore LMN is the plane required.

Corollary. *To determine the traces of a plane which shall bisect the angle between two planes.*

Find the angle θ between the two planes by Prob. XXI. Next draw a third plane containing the common section of the two given planes and making an angle $\dfrac{\theta}{2}$ with either of them.

PROBLEM XXIII. Figs. 51 and 52.

To determine the inclination of a straight line to a plane.

Let LMN be the traces of a plane, and ab, $a'b'$ the projections of a line, it is required to find the angle which AB makes with LMN.

Construction. From any point A on the given line draw AC perpendicular to the given plane (Prob. XV.), and find the angle contained by AB and $AC = bA_1c$ (Prob. VIII.). The angle θ which is the complement of bA_1c will be the angle required.

Proof. Let P and Q be the points in which AB and AC respectively meet LMN.

Then PQ is the projection of AP on the given plane, and consequently the angle APQ is the angle required (Def. 13). But AQP is a right angle (Def. 5). Therefore $APQ = 90° - PAQ = \theta$.

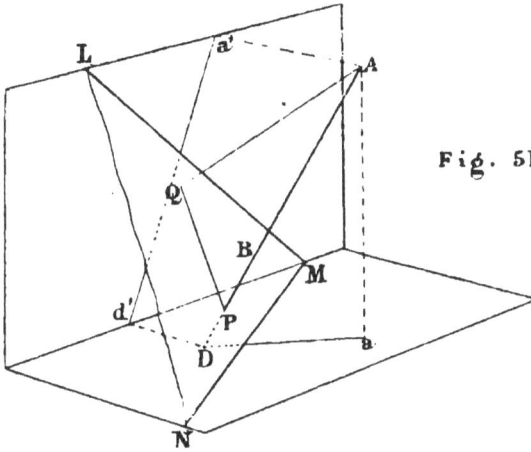

Fig. 51.

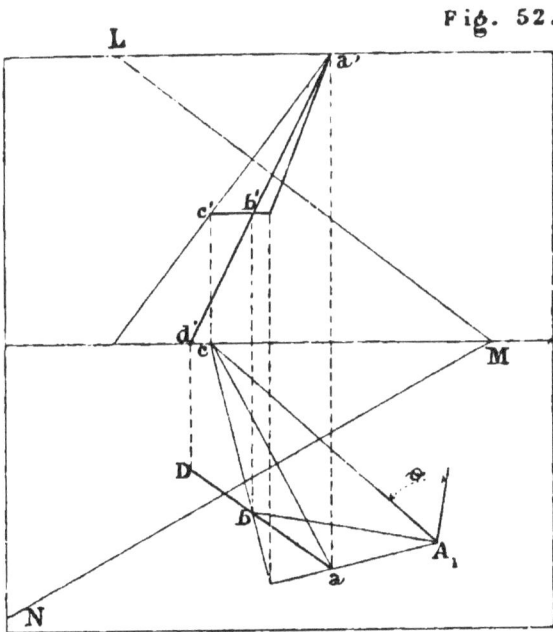

Fig. 52.

PROBLEM XXIV.   Figs. 53 and 54.

Given the traces of a plane and one of the projections of a point in the plane, to find the position of the point when the plane is rabatted on the horizontal plane.

Let LMN be the given plane and p the horizontal projection of the point; it is required to find the position of P when LMN is turned about MN till it coincides with the horizontal plane.

Construction. Draw the projections of any line containing P and lying in LMN (in the figure the projections of the horizontal line PQ are drawn), and so determine p', the vertical projection of P.

Next, draw pR at right angles to MN, and make RP_1 on the line produced equal to RP (Prob. IV.).

P_1 is the point required.

Proof. The line PR is at right angles to MN (Theor. XVIII.) ; and as the plane LMN turns about MN, the line PR will remain at right angles to MN and take the position P_1R (Theor. III. Cor. 2).

But $PR = P_1R$.

Therefore P_1 is the point required.

PROBLEM XXV.   Figs. 53 and 54.

Given the horizontal trace of a plane and the position of a point in the plane when it is rabatted on the horizontal; to find the projections of the point when the plane is raised into any given position.

Let MN be the horizontal trace of the plane and P_1 the position of the point when the plane rests on the horizontal ; it is required to find the projections of P when the plane is raised to the position LMN.

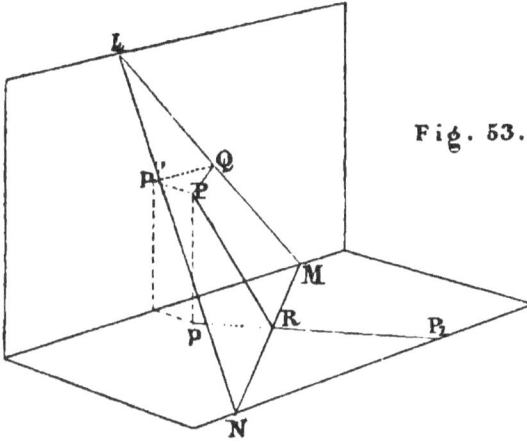

Fig. 53.

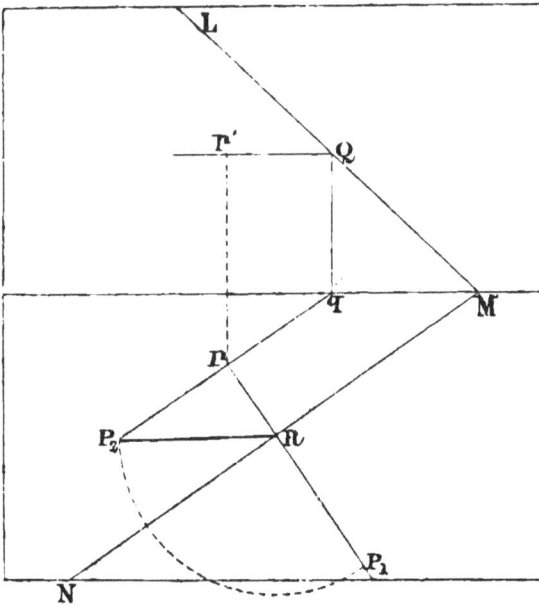

Fig. 54.

Construction. Draw $P_1 p$ at right angles to MN, and construct the right-angled triangle $RP_2 p$ having the hypotenuse RP_2 equal to RP_1, and the angle $P_2 Rp$ equal to the inclination of LMN to the horizontal (Prob. XIX.). p is the horizontal projection required, and p', at a distance from the ground line equal to $P_2 p$, the vertical projection.

Proof. When the plane LMN is turned about MN, the point P_1 will take such a position P, that PR will be at right angles to MN (Theor. III. Cor. 2); also, since $PR = P_1 R$ $= P_2 R$, and the angle $P_2 Rp = PRp$, the triangle PpR will be equal to $P_2 pR$.

Therefore p and p' are the points required.

PROBLEM XXVI. Figs. 55 and 56.

From a given point to draw a line to make a given angle with a given line.

Let A be the given point and BC the given line; it is required to draw from A a line to make a given angle θ with BC.

Construction. Find the traces LM and MN of the plane containing A, B, and C (Prob. X.).

Let the plane LMN be rabatted on the horizontal plane, and find the position of the point B, viz. B_1 (this is done in the figure by drawing bB_1 at right angles to MN and making MB_1 equal to MB) and the position A_1 of A (Prob. XXIV.). Join $B_1 C$, and from A_1 draw $A_1 D_1$, making the angle $A_1 D_1 C = \theta$.

Next find the projections dd' of D_1 when the plane LMN is raised to its former position (Prob. XXV.).

AD is the line required.

Proof. If the plane LMN were turned about MN till it coincided with the horizontal plane, it is evident from the construction that the three points A, C and D would coincide respectively with A_1, C and D_1 (Prob. XXIV.). Therefore $ADC = \theta$.

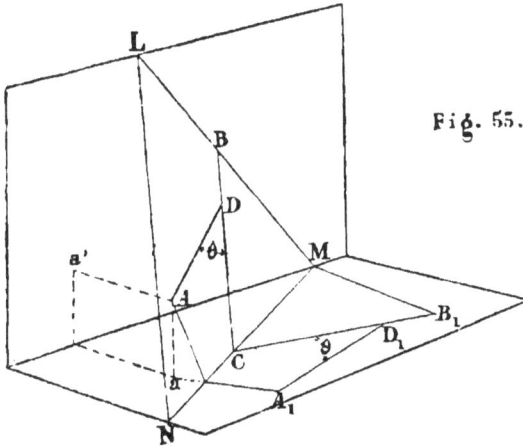

Fig. 55.

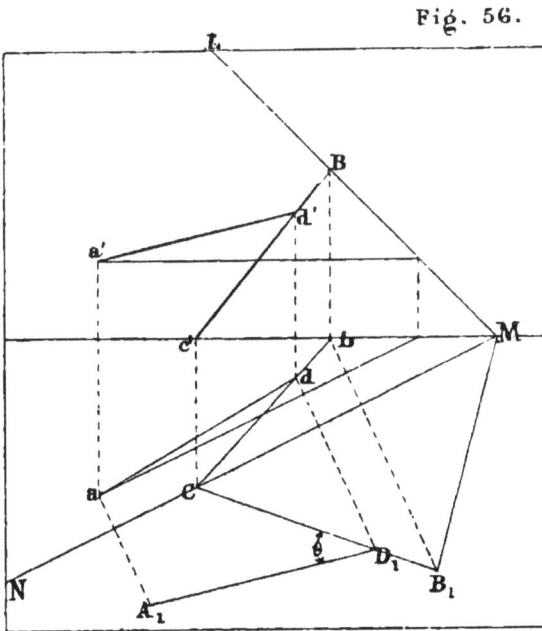

Fig. 56.

Exercises.

1. *A* and *B* are the horizontal and vertical traces respectively of a straight line; a' and b the projections of *A* and *B*; $Aa' = 2''$, $a'b = 2\frac{1}{2}''$, $Bb = 3''$. Draw the projections of *AB*, and determine its length.

2. Mark the projections of four points, one in each of the dihedral angles contained by the planes of projection, each point being $1''$ distant from the horizontal plane and $2''$ from the vertical plane. Find the traces and the length between the points of every line that can be drawn through two of them.

3. The heights of two points *A* and *B* above the horizontal plane of projection are $1''$ and $2\frac{1}{2}''$ respectively, and *ab*, which is $2''$ long, makes an angle of $60°$ with *xy*. Find the inclination of *AB* to each plane of projection and mark the projections of a point on it $2''$ from *A*.

4. Represent a line $2''$ long in each of the following positions.

(1) Horizontal and inclined at $30°$ to the vertical plane of projection.

(2) Inclined at $30°$ to the horizontal plane, and its horizontal projection making an angle of $40°$ with *xy*.

(3) Making an angle of $45°$ with the horizontal and an angle of $30°$ with the vertical plane of projection.

5. The vertices of a triangle *ABC* are respectively $1''$, $2\frac{1}{2}''$ and $3''$ above the horizontal plane of projection, and its plan is an equilateral triangle of $1\frac{1}{2}''$ side.

(1) Find the horizontal traces of the sides of the triangle.

(2) Find their inclinations to the horizontal plane.

(3) Construct a triangle equal in all respects to *ABC*.

(4) Find the projections of the centre of the circle passing through *A, B* and *C*.

6. Draw the plan of an equilateral triangle of 2″ side when its vertices are 1″, 1½″ and 2¼″ respectively above the horizontal plane of projection.

7. Draw the plan of an equilateral triangle of 2″ side when two of its sides are inclined at angles of 30° and 45° respectively to the horizon.

8. Two lines AB, AC, at right angles to one another, are inclined at angles of 40° and 30° respectively to the horizontal plane of projection. Draw their horizontal projections, and their vertical projection on a plane containing AB.

9. Find the horizontal trace of the plane of each pair of given lines in the last three exercises.

10. Take xy making an angle of 45° with ab of Ex. 5 and determine the traces of the plane of the triangle ABC.

11. Represent a plane in each of the following cases :—

(1) Perpendicular to the two planes of projection.

(2) Perpendicular to the vertical plane of projection and inclined at 50° to the horizontal.

(3) Parallel to the ground line, and inclined at 50° to the horizontal.

12. The horizontal and vertical traces of a plane make angles of 40° and 50° respectively with xy; find a point in the plane 1″ from each plane of projection, and draw a line passing through that point, lying in the plane, and inclined at 30° to the horizontal plane.

13. Determine the real angle between the traces of the plane in the last exercise.

14. The common section of two planes is 4″ long between its traces, is inclined 30° to the vertical plane of projection and 40° to the horizontal; the horizontal traces of the planes make angles of 70° with xy: draw both traces of each plane. Also draw a vertical plane making an angle of 45° with xy, and determine the angle contained by the lines in which it cuts the two former planes.

15. There are four lines AB, CD, EF, GH, parallel to xy, they are $1''$, $1\frac{1}{4}''$, $2''$ and $2\frac{3}{4}''$ respectively above the horizontal plane of projection, and $3''$, $2\frac{1}{4}''$, $1\frac{3}{4}''$ and $1''$ in front of the vertical plane of projection: determine the traces of the planes $ABEF$, $CDGH$, and the projections of their common section.

16. Through a point $2''$ above the horizontal plane of projection and $1''$ from the vertical plane, draw three planes perpendicular to one another, none of them being parallel to the ground line.

17. There are two triangles LMO, MNO, on a common base MO;

$$LM = 3'', \ LO = 2'', \ MO = 3\frac{1}{2}'', \ MN = 2\frac{1}{2}'', \ ON = 4'';$$

LM, LO are the vertical traces and MN, NO the horizontal traces of two planes.

(1) Find the inclinations of the two planes to the planes of projection and to one another.

(2) Determine two planes bisecting the angles between the given planes.

18. Draw the traces of a plane inclined at $80°$ to the vertical plane of projection and $40°$ to the horizontal.

19. In the plane of the preceding problem take a point $1''$ from each plane of projection and determine the traces of a plane passing through that point, inclined at $60°$ to the former plane and cutting it in a line inclined to the horizontal at $30°$.

20. A plane mirror is inclined at $40°$ to the horizontal plane of projection; a ray of light passes through a point P before reflection and a point Q after reflection; pq, the horizontal projection of PQ, makes an angle of $60°$ with the horizontal trace of the plane of the mirror, and is cut by it in a point r, so that

$$pr = 2'', \ rq = 3''; \ Pp = 2\frac{1}{2}'', \ Qq = 4'':$$

find the distance of the point of reflection from P and Q, and the angle of reflection.

CHAPTER III.

DEFINITION 1. When three or more planes meet in one point only, they form at that point a *solid angle*.

The lines of intersection of the planes are called *edges*, and the parts of the planes between the edges the *sides* or *faces*.

DEF. 2. A *polyhedron* is a solid figure contained by plane figures, which are called its *faces*.

If it have four faces only it is called a *tetrahedron*; if six, a *hexahedron*; if eight, an *octahedron*; if twelve, a *dodecahedron*; if twenty, an *icosahedron*.

The points of intersection of the edges of a polyhedron are called *vertices*, and the straight line joining any two vertices, not lying in the same face, is called a *diagonal* of the solid.

DEF. 3. A *polyhedron* is said to be *regular* when its faces are regular plane figures equal and similar to one another.

There are only five regular polyhedrons; those named in Def. 2.

DEF. 4. A *pyramid* is a polyhedron contained by a plane figure, called its base, and by three or more triangles meeting in a point without the plane of the base. This point is called the vertex of the pyramid.

DEF. 5. A prism is a polyhedron having any number of faces, two of which, called the bases, are equal and similar, and so placed as to have their corresponding sides parallel, and the rest parallelograms.

DEF. 6. Pyramids and prisms are called *triangular, quadrilateral, pentagonal*, &c., according as the figures of their bases are triangles, quadrilaterals, pentagons, &c.

DEF. 7. The *altitude or height of a pyramid* is the perpendicular distance from the vertex to the plane of the base: and the *altitude of a prism* is the perpendicular distance between the planes of its bases.

DEF. 8. A *regular pyramid* is that which has a regular figure for its base, and the vertex on the normal to the base through the centre. That normal when it contains the vertex is called the *axis* of the pyramid.

DEF. 9. A *regular prism* is that which has its bases regular figures. When the line joining the centres of the bases, called the axis, is perpendicular to them the prism is said to be *right:* any other is *oblique.*

DEF. 10. A parallelopiped is a prism of which the bases are parallelograms.

DEF. 11. A *rectangular parallelopiped* is one which has its bases and the other faces all rectangles.

DEF. 12. A *cube* is a rectangular parallelopiped of which all the faces are squares.

PROBLEM I. Fig. 57.´

Draw the plan of a square of $1\frac{1}{4}''$ *side when its plane is inclined at* 45° *and a side AB inclined at* 36°.

Draw Aa' at right angles to *xy*, and $a'P$ making an angle of 45° with *xy*; $Aa'P$ are the traces of a plane having the required inclination of 45° and perpendicular to the vertical plane of projection.

Take any point P in the vertical trace of the plane and draw PQ making an angle of 36° with *xy*. Let the right-angled triangle PpQ be conceived to turn about the axis Pp, as in Problem VII. Chap. II., till the point Q comes to A in

the horizontal trace of the plane. A and P are the traces,
and consequently Ap, $a'P$ the projections, of a line lying in
the plane $Aa'P$ and having an inclination of $36°$.

Now let the plane $Aa'P$ be rabatted on the horizontal
plane; the line AP will obviously come into the position
AP_1. Make $AB_1 = 1\frac{1}{4}''$ and on AB_1 construct the square
$AB_1C_1D_1$ which will be the rabatment of the square required.
When the plane containing the square is raised to its former
position the points b_1', c_1', d_1' come to b', c', d', which are therefore the vertical projections of A, B, C, three angular points
of the square, and $Abcd$ is the plan required.

It will be observed in this Problem how much the con-
structions are simplified by taking the plane containing the
figure perpendicular to the vertical plane of projection.

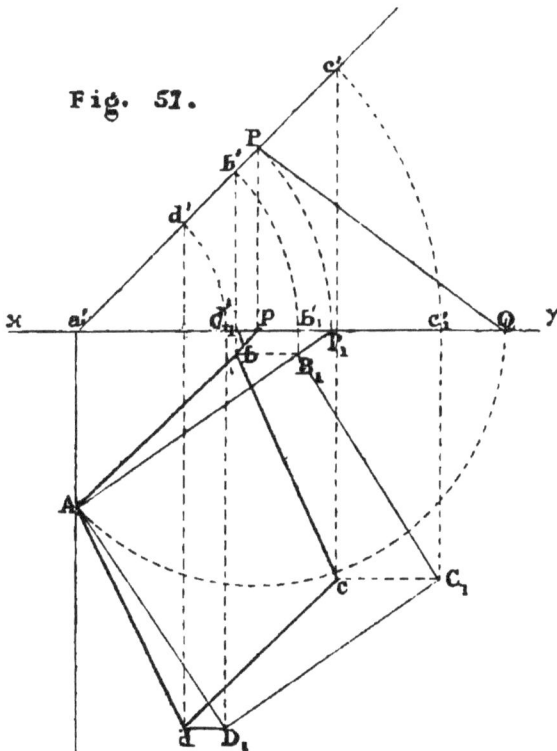

Fig. 51.

PROBLEM II. Fig. 58.

The side of a regular hexagon is 1″ *; its plane is inclined at* 50° *; and two adjacent angular points A and B are* 0·4″ *and* 1″ *respectively above the horizontal plane of projection. Draw its plan.*

As in Problem I. draw $Ss'a'$, the traces of a plane inclined at 50° and perpendicular to the vertical plane of projection. Find the two points a' and b' on the vertical trace of the plane at a distance 0·4″ and 1″ respectively from xy: a', b' are the traces of two horizontal lines in the plane $Ss'a'$. Now let the plane be rabatted on the horizontal plane and draw the two lines $a_1'm$ and $b_1'n$ parallel to Ss'; these are the rabatments of the two horizontal lines through a' and b'.

Next draw A_1B_1 an inch long and having one extremity on each of the lines $a_1'm$ and $b_1'n$; on A_1B_1 construct the regular hexagon $A_1B_1C_1D_1E_1F_1$.

Now let the plane be raised to its former position as in the last problem; $abcdef$ is the plan of the hexagon.

As a check on the accuracy of the construction it may be noticed that if any line of the hexagon $A_1B_1C_1...$ be produced to meet Ss', the horizontal trace of the plane, as A_1B_1 and E_1F_1, the projections ab and ef of these lines must meet Ss' in the same points S and T respectively, for as the plane containing the regular hexagon turns about Ss' as an axis every point in that line remains fixed.

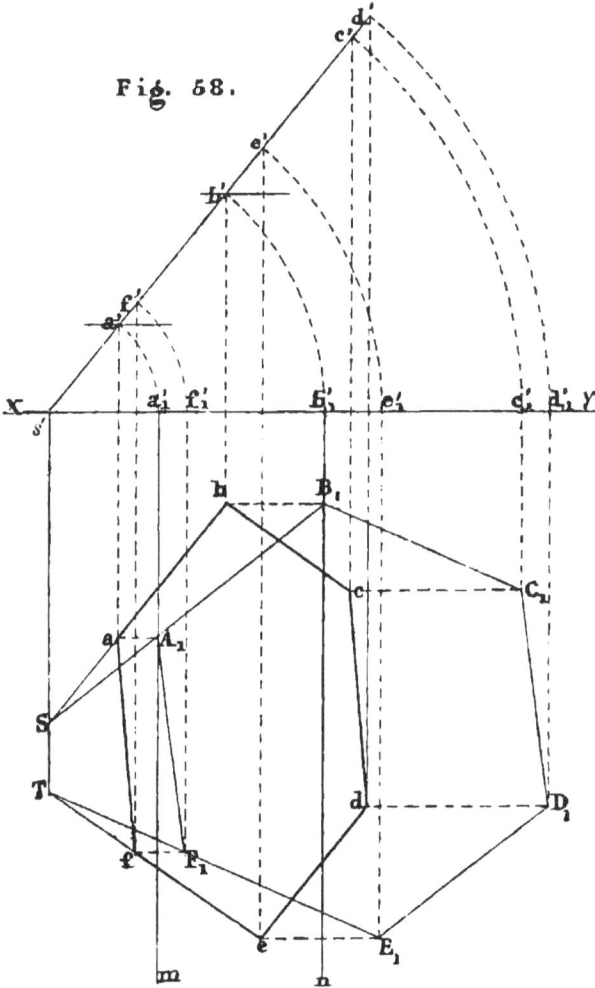

Fig. 58.

PROBLEM III. Fig. 59.

Two adjacent sides, AB *and* AD, *of a parallelogram are* 1″ *and* 1¼″ *respectively, and the contained angle* 45⁰. *Find the plan of the parallelogram when* AB *is inclined at* 30° *and* AD *inclined at* 50″. *Draw also an elevation on a vertical plane containing* AB, *and another on a plane making an angle of* 80″ *with* ab.

Taking any ground line *xy* and any point *A* find by Problem IX. Chap. II. the projections of two lines intersecting at *A*, containing an angle of 45″, and having inclinations of 30⁰ and 50⁰ respectively. *A F* is one of these lines, in the vertical plane of projection, and aE_1, Ae_1' the projections of the other. Set off on *A F* the distance *A B* = 1″ and on *A E,* the first position of the second line, the distance $AD_1 = 1\frac{1}{4}″$. *ab* and *ad* (=ad_1) are the horizontal projections of the two sides of the given parallelogram in the position required. The figure is completed by drawing *bc* and *cd* parallel to *ad* and *ab* respectively. *A Bc'd'* is the elevation on the plane containing *A B*.

To find an elevation on a plane making an angle of 80⁰ with *ab*. Draw *x'y'* making an angle of 80° with *ab* for a new ground line, and draw *aa″, bb″, cc″, dd″* at right angles to *x'y'*, making the distances of *a″, b″, c″, d″* from *x'y'* equal respectively to the distances of *A, B, c', d'* from *xy* ; for these distances are the heights of *A, B, C* and *D* respectively above the horizontal plane of projection. *a″b″c″d″* is the elevation required.

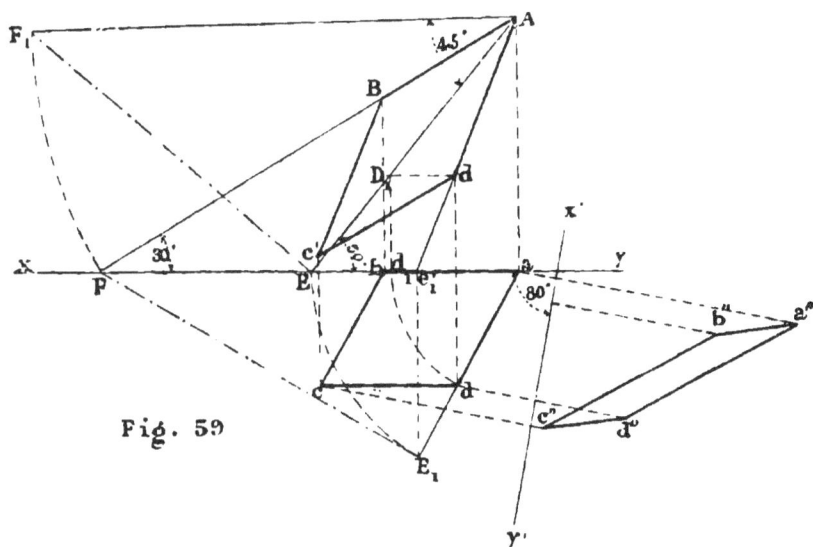

Fig. 59

PROBLEM IV. Fig. 60.

Three consecutive angular points of a regular pentagon, A, B and C are 0·7″, 1·2″ and 0·9″ respectively above the horizontal plane and the side of the pentagon is 1″. Determine its horizontal projection.

Draw a line Bb at right angles to the ground line; and set off on it bm, bn, bB, equal respectively to 0·7″, 0·9″ and 1·2″, the heights of the three points. Draw the horizontal lines mA_1, nC, and from B describe an arc of a circle with 1″ radius cutting them at A_1 and C. Join BA_1 and BC and produce the lines to meet xy at F and G. Then $BFb = BA_1m$ and $BGb = BCn$ are the inclinations of the two sides of the pentagon meeting at B. From the inclinations of the two lines and the contained angle, *i.e.* 108°, find their projections bG, bF_1......(Prob. IX. Ch. II.).

GF_1 is obviously the trace of the plane containing the two sides of the pentagon BA and BC. Let the plane be rabatted about GF_1 as in Prob. XXIV. Ch. II.

The point B comes to B_1, A to A_2, and C to C_1. The pentagon being now constructed on the horizontal plane, the plane containing the figure may be returned to its former position and the projections of D_1 and E_1 found as in Prob. XXV.

Remarks. The points e and d are found conveniently in this case by observing that A_2E_1 and ae meet GF_1 at the same point which is the horizontal trace of AE; and so of C_1D_1 and cd; but as the perpendicular from D_1 on GF_1 cuts cd rather obliquely for accurately determining d, the diagonal bd has been drawn.

In every problem there are generally several ways of determining points, and it is for the student to choose that which is the most simple and accurate. The different methods serve as checks on one another, and it is well to observe these. For instance in this Problem $BG = B_1G$, $BF = B_1F_1$ and the angle GB_1F_1 should be 108°; also B_1 should be on the perpendicular from b on GF.

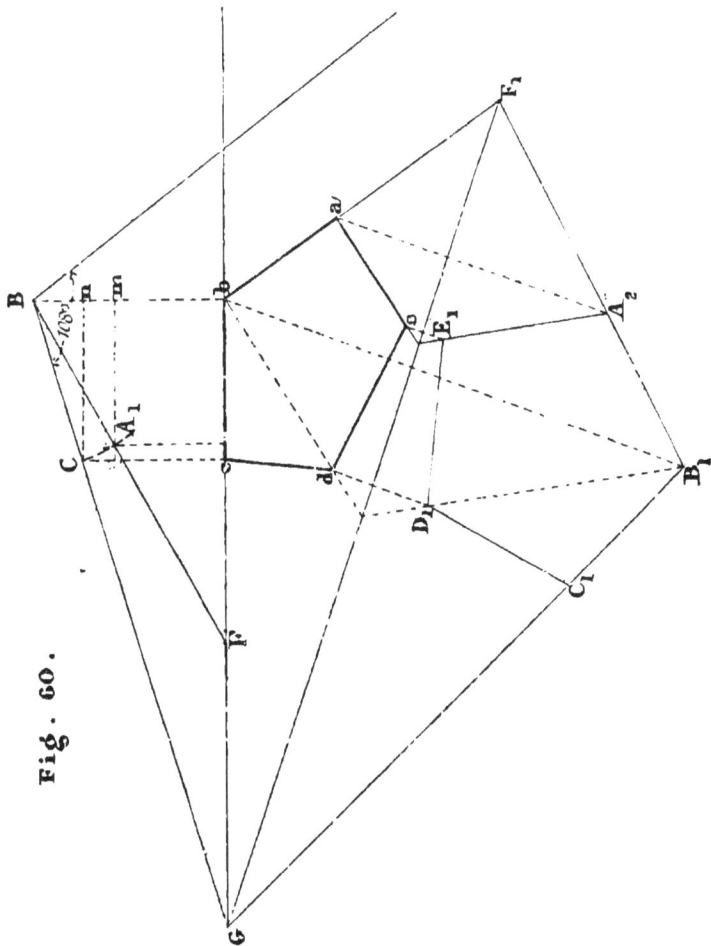

Fig. 60.

Problem V.   Fig. 61.

Find the horizontal projection of a circle of 2″ diameter when its plane is inclined at 50°.

Draw the traces LMN of a plane inclined at 50°, and taking a point cc' in that plane as the centre of the circle draw ab through c parallel to MN, and make ac, bc each equal to the radius of the circle.

Now suppose the circle to revolve about the horizontal diameter AB till it is parallel to the horizontal plane; its vertical projection is the straight line $d_1'e_1'$ and its horizontal projection the circle ad_1be_1...Theor. xv. Cor. 1.

Take any number of points d_1d_1', e_1e_1', f_1f_1', &c., in the projection of the circle in this position, and let it be returned to its former position by revolving about AB; the points d_1', f_1', &c. come to d', f'', &c., and their horizontal projections, lying on the lines d_1c, f_1r, &c. perpendicular to ab, are d, f, g, e, &c. The curve traced through these points is the projection required.

In the figure $od'c'$, $d'c'$: oc' :: $f'c'$: pc' :: $g'c'$: qc' ;

that is, d_1c : dc :: f_1r : fr :: g_1s : gs.

But this is the well-known relation of an ellipse to its auxiliary circle; hence—

If the plane of a circle be inclined to any other plane its projection on that plane shall be an ellipse of which the major axis shall be the projection of the diameter of the circle parallel to the plane, and therefore equal to it, and the minor axis the projection of the diameter of the circle at right angles to the former one, being that diameter which has the greatest inclination to the given plane.

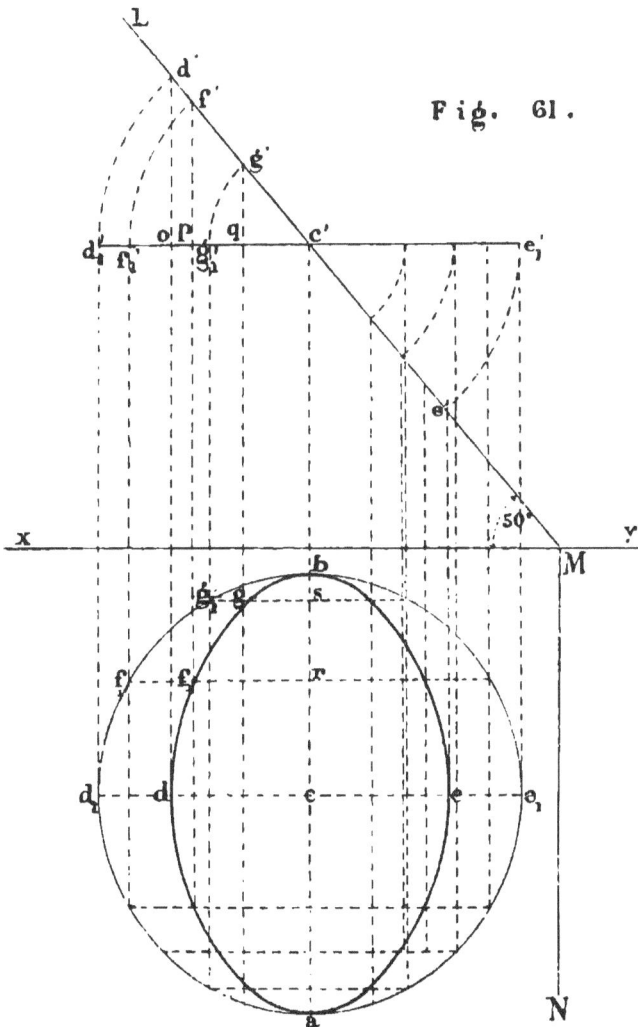

Fig. 61.

PROBLEM VI. Figs. 62, 63, 64.

The side of the base of a regular hexagonal pyramid is $\frac{3}{4}''$ and its height $2''$. Draw its projections in the following positions:—

(1) *When standing with its base on the horizontal plane, and one side of the base at right angles to the ground line.*

(2) *When resting on one edge of the base so that the axis is inclined at $45°$, the ground line being at right angles to the edge on which it rests.*

(3) *When resting with one of the triangular faces on the horizontal plane, and the axis parallel to the vertical plane.*

In the last position show a section by a vertical plane which cuts the axis at its middle point and makes an angle of $60°$ with the vertical plane of projection.

1st position. Draw the regular hexagon $ABCDEF$ with a side of $\frac{3}{4}''$ in such a position that a side BC is at right angles to xy; and join v, the centre, with each of the angular points; this is the plan of the pyramid. Draw vv' at right angles to xy, and make $a'v' = 2''$; join v' with b' and f : this is the elevation required.

2nd position. To obtain the projections of the pyramid in the second position either of two methods may be used. The horizontal plane of projection may be supposed to change its position while the pyramid remains fixed; or the planes of projection remaining as before the position of the solid may be changed relatively to them. The former is generally the shortest method, but to prevent confusion of the figures the latter has been adopted in this case. The pyramid is supposed to move along the horizontal plane, each point moving parallel to the vertical plane, and then to be turned about EF till the axis makes an angle of $45°$ with the horizontal plane, that is till $a'v'$, its projection, makes $45°$ with xy, the axis being parallel to the vertical plane. The remainder of the construction will be evident from the figure.

3rd position. This construction is similar to that used for the 2nd position, the pyramid being turned about EF till the face EFV coincides with the horizontal plane.

Section. LMN is the cutting plane. The way in which the vertical projection of the section is obtained will be evident from the figure.

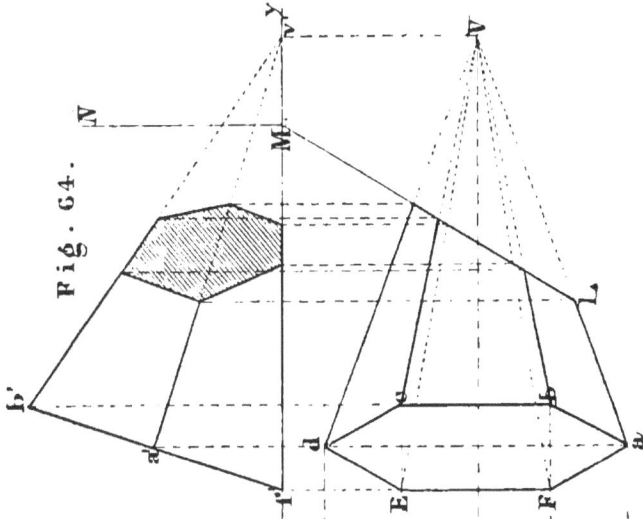

Fig. 64.

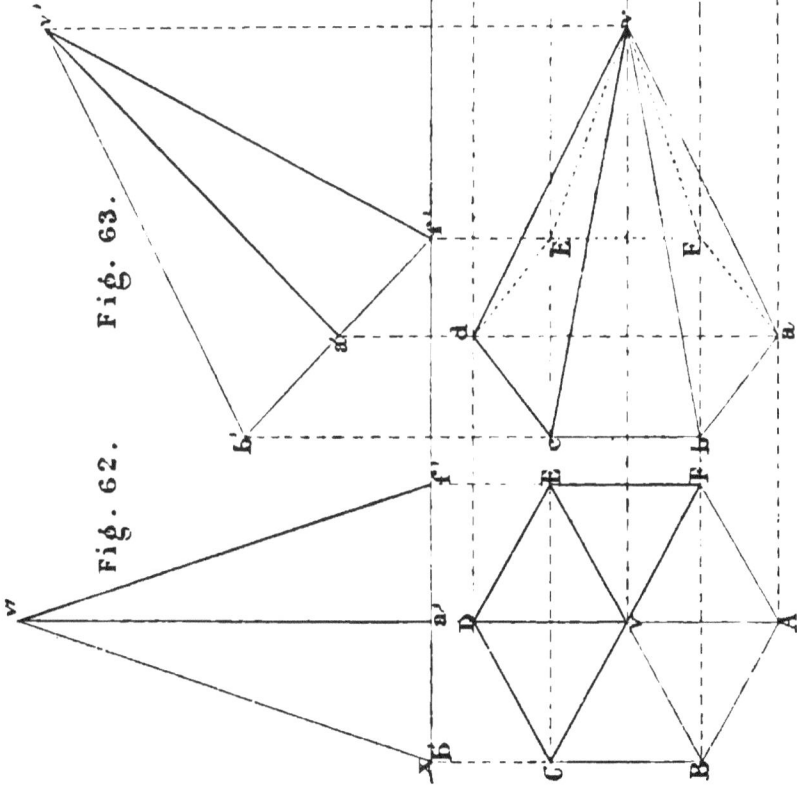

Fig. 63.

Fig. 62.

PROBLEM VII.   Fig. 65.

Draw the projections of a cube of 1¼″ *edge when resting with one edge on the horizontal plane, that edge making an angle of* 30⁰ *with the ground line, and a plane containing it inclined at* 40°.

There are two distinct steps in this problem. First draw the square *a'b'c'd'* having one side *a'b'* making an angle of 40° with *xy*; this is the elevation of a cube having one face parallel to the vertical plane and a face at right angles to the former one inclined at 40″. Next draw the plan as in the figure, the edge *be* and those parallel to it being equal to the edge of the cube since they are parallel to the horizontal plane.

The further condition of the problem is obtained by drawing a new ground line *x₁y₁* making an angle of 30° with the edge on which the cube rests, and drawing the elevation *a″b″c″* &c....by making the distance of each point from the ground line equal to its height above the horizontal plane of projection. These heights are found from the elevation *a'b'c'd'*.

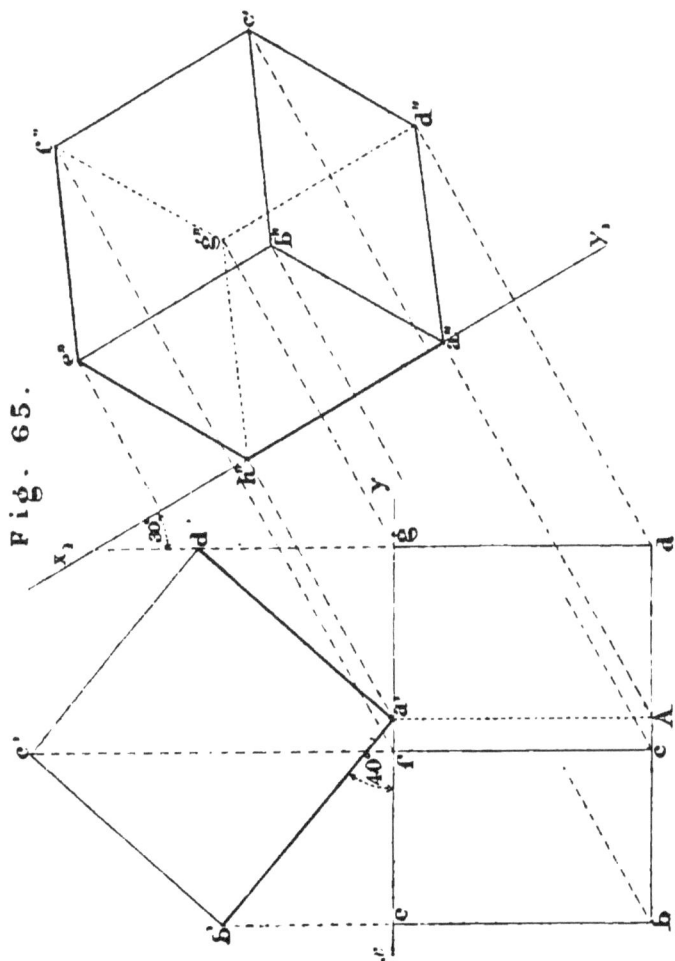

Fig. 65.

PROBLEM VIII. Fig. 66.

Draw the projections of a regular octahedron of 1″ edge in the following positions :—

(1) *Plan and elevation when a diagonal is vertical, and the vertical plane of projection parallel to one of the horizontal edges.*

(2) *Plan, when one of its faces is horizontal.*

A regular octahedron has eight equal faces which are equilateral triangles; but perhaps the easiest way to realize its form to the mind is to conceive it as made up of two pyramids having a common base, which is a square, and the triangular faces equilateral.

When a diagonal is vertical the plan of the octahedron is the same as that of a square pyramid whose axis is vertical.

(1) Draw the square *bcde* having a side of 1″, and the two diagonals *bd* and *ec*; that is the plan of the octahedron in the required position.

To find the elevation, draw *aa′* at right angles to *xy*, and make *f′a′* equal to *bd*—all diagonals of the solid being equal—bisect *f′a′*, and draw *b′e′* at right angles to it through the point of bisection, to meet *be* and *cd* produced : *a′b′e′f′* is the vertical projection required.

(2) When one of its faces is horizontal the vertical projection of that face is a straight line parallel to the ground line. Draw, then, *x₁y₁* parallel to *b′f′* for a new ground line, and find the projection *a″b″c″f″d″e″*, by making the distances of these points from *x₁y₁* equal respectively to the distances of *abcfde* from *xy*; these being the actual distances of *ABCFDE* from the vertical plane of projection which is the same for both plans. It may be observed that this construction is equivalent to turning the solid about any horizontal line perpendicular to the vertical plane till the face *BEF* is horizontal.

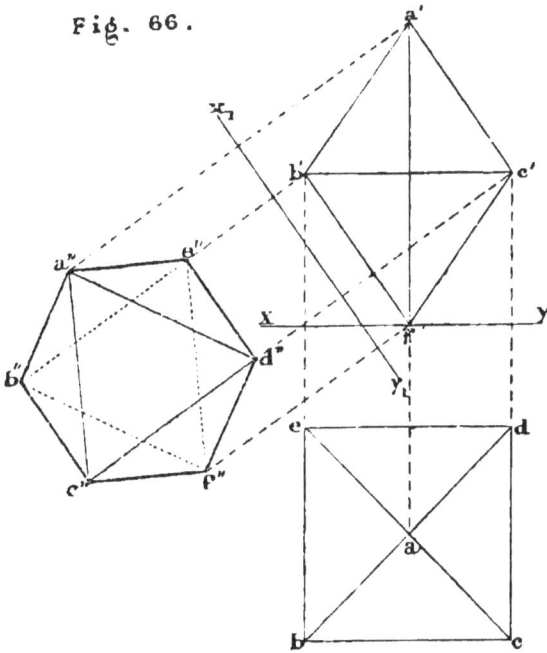

Fig. 66.

PROBLEM IX. Fig. 67.

The base ABC of a regular tetrahedron is inclined at 50°, and the edge AB of the base at 30°. Draw its plan ; and an elevation on a plane parallel to AB. Edge = 1¼".

The tetrahedron is a pyramid of which the base and the other three faces are equilateral triangles.

Draw the projections Abc, $a'b'c'$, of an equilateral triangle whose plane is inclined at 50° and the side AB at 30°, as in Prob. I. Through the centre of ABC draw a line perpendicular to its plane and equal to the height of the tetrahedron. That height is equal to one side of a right-angled triangle of which Ad_1 is the base, and an edge of the solid, as AB_1, the hypotenuse ; it is parallel to the vertical plane and is equal to its projection, $e'd'$. This gives dd' which is the fourth vertex of the tetrahedron. The rest of the construction is evident from the figure. $a''b''c''d''$ is the required elevation.

Fig. 67

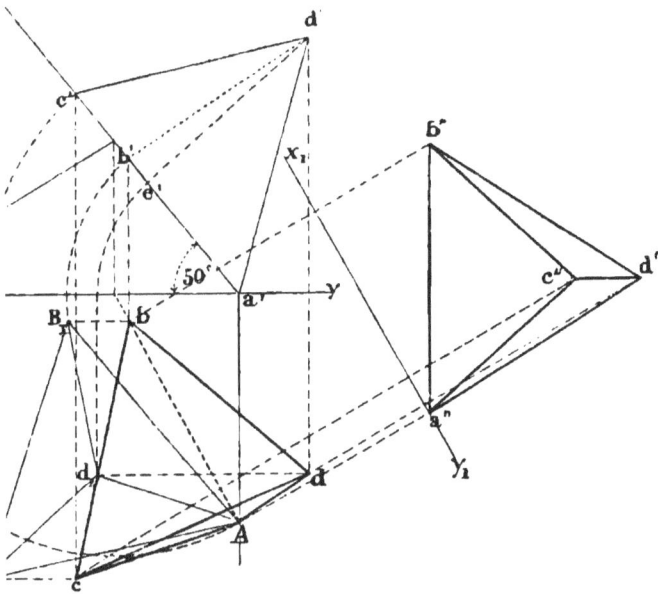

PROBLEM X. Fig. 68.

The extremities of the edge of a cube are $1''$ *and* $1\frac{1}{2}''$ *respectively above the horizontal plane, and a face containing that edge is inclined at* 45°. *Draw the plan of the cube.* $Edge = 1\cdot2''$.

Find, as in Problem II., the projections $abcd$, $a'b'c'd'$ of a square of $1\cdot2''$ side whose plane is inclined at 45°, and the points A and B, $1\frac{1}{2}''$ and $1''$ respectively above the horizontal plane. $abcd$ is the plan of one face of the cube. Through aa' draw a line AE perpendicular to LMN, the plane of $ABCD$, by Prob. XV. Ch. II., and make it equal to the edge of the cube; ae is the horizontal projection of that line. In a similar way are found f, g and h, the projections of the other angular points of the cube.

Fig. 68.

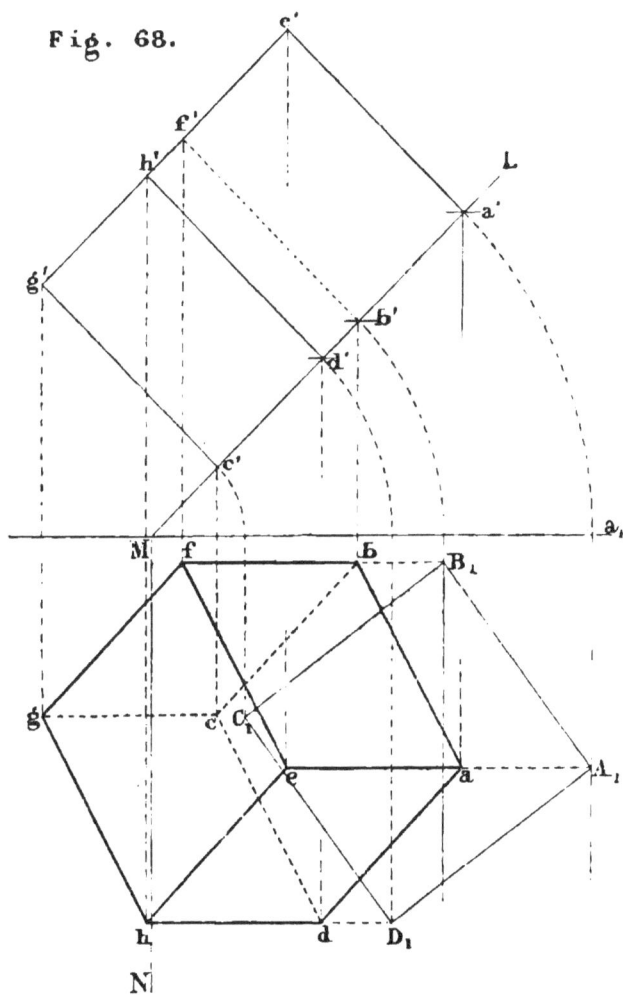

PROBLEM XI.   Fig. 69.

An octahedron of $1\cdot5''$ edge has two of its diagonals inclined at $30°$ and $50°$ respectively.   Draw its plan.

Find the projections of two lines oP and oQ_1 at right angles to one another, and having the required inclinations of $30°$ and $50°$ respectively by Prob. IX. Ch. II.   Next find the projections a and b of two points on those lines, whose distance from O is equal to half the diagonal of a square of $1\cdot5''$ side.   Produce ao to c making $oc = oa$, and bo to d making $od = ob$.

$abcd$ is the projection of a square of $1\cdot5''$ side, and having its diagonals inclined at $30°$ and $50°$ respectively.

Draw the projections of a line passing through O and perpendicular to the plane of $abcd$, that is to the plane OPQ_1. Prob. XV. Chap. II.

oT, Ot' are the projections of the line.

Find the projections e and f of two points on that line whose distance from O is equal to half the diagonal of the square $ABCD$.... Prob. V. Chap. II.

$abcdef$ is the plan of the octahedron.

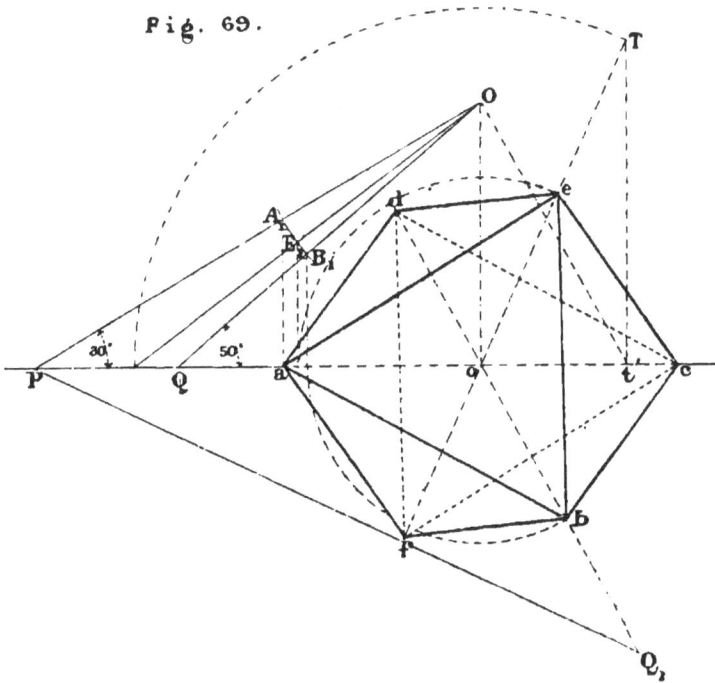

Fig. 69.

Note. The triangle *aeb* should be dotted, and *cdf* drawn in full lines.

PROBLEM XII.   Fig. 70.

The base of a rectangular parallelopiped is a square of 1″ side, and its length is 2″. Draw its plan, when three consecutive angular points of the base A, B, C are ·3″, ·75″ and 1·25″ respectively above the horizontal plane.

Draw the horizontal lines mA_2 and nB at the distances ·3″ and ·75″ respectively from xy; and make cC=1·25″. Make $CB = 1″$, the side of the square; and CA_2 = diagonal of the square. Next find the horizontal projections of the two lines CA_2 and CB when they have the inclinations CA_2m and CBn, and the angle between them, $BCA_1 = 45^0$. ca and cb are these projections; and the parallelogram $abcd$ is the projection of the square end of the solid.

Through the point C draw a line perpendicular to the plane of $ABCD$, and find a point on it 2″ from C.... Probs XV. and V. Chap. II.

cg is the projection of that line, which is one of the edges of the parallelopiped.

ae, bf, dh are equal and parallel to cg, and $abcdefgh$ is the projection required.

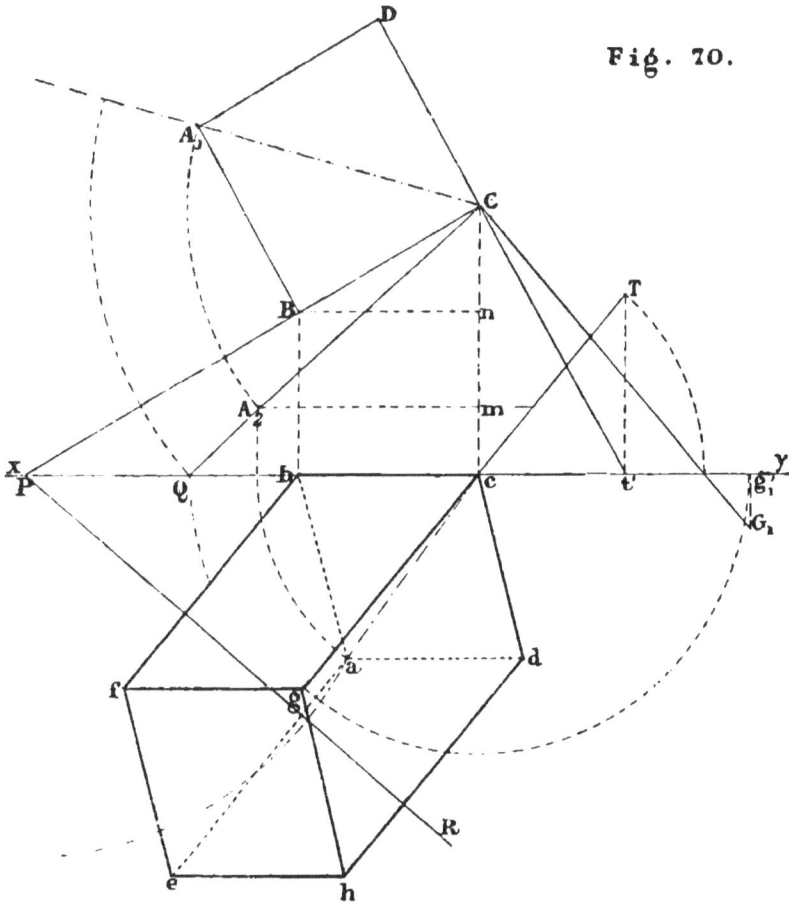

Fig. 70.

The Trihedral Angle.

In a trihedral angle there are three plane angles or *faces* and three dihedral angles. When any three of these six parts are known the other three may be determined. There are therefore six different cases. Let the three dihedral angles be denoted by A, B, C; and the plane angles respectively opposite to them by a, β, γ.

The data of the six cases are;

(1) a, β, γ ;   (4) A, B, C;
(2) a, β, C;   (5) A, B, γ;
(3) a, β, A ;   (6) A, B, a.

That is, (1) the three faces ; (2) two faces and their inclination to one another ; (3) two faces and the dihedral angle opposite to one of them ; (4) the three dihedral angles ; (5) a face and two adjacent dihedral angles; (6) two dihedral angles and the face opposite one of them. It will afterwards be shown that the last three cases may be reduced to the first three.

Problem I.   Figs. 71 and 72.

Given the three faces of a trihedral angle, to find the three dihedral angles.

Let the three angles a, β and γ be constructed as in fig. 72 on the same plane, which for convenience may be supposed horizontal; and let the points P_1 and P_2 be at equal distances from S.

Now if the triangles SP_1Q and SP_2R were made to revolve about SQ and SR respectively till the two lines SP_1, SP_2 coincided, it is evident a trihedral angle would be formed having the three given faces.

But as the triangle SP_1Q revolves about SQ the locus of the projection of the point P_1 is P_1M at right angles to SQ.

Similarly the locus of the projection of P_2 is P_2N at right angles to SN; and as $SP_1 = SP_2$, the points P_1 and P_2 must meet at P when the lines coincide.

Therefore p is the projection of a point on the third edge of the solid angle ; and Sp the projection of that edge.

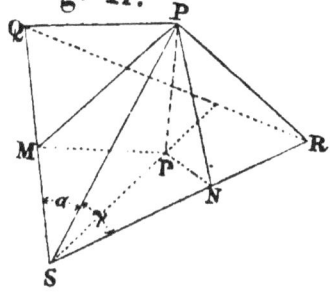

Fig. 71.

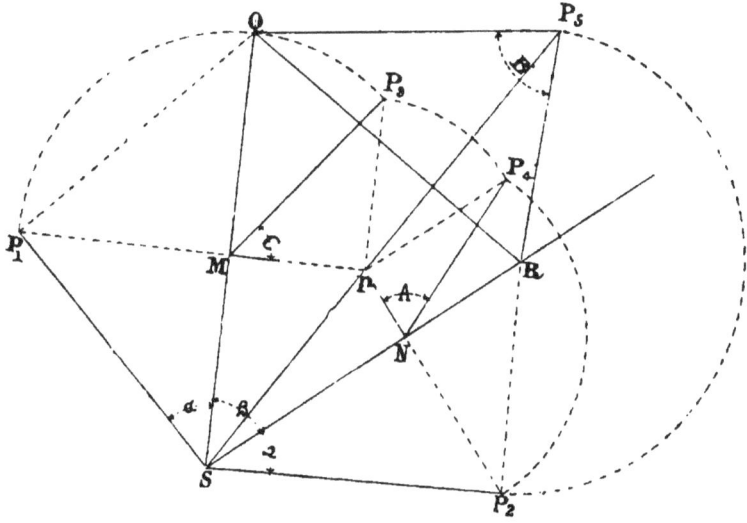

Fig. 72.

The angle C is the angle contained by the two lines PM and Mp; so that to find C it is only necessary to construct the right-angled triangle P_3Mp, having the hypotenuse $MP_3 = MP_1 = MP$.

The angle A is found similarly by constructing the right-angled triangle P_4Np, having $P_4N = NP_2 = NP$.

The angle B is the angle contained by PQ and PR, which are at right angles to PS. But these two lines are equal respectively to P_1Q and P_2R, at right angles to SP_1 and SP_2.

Let the triangle QP_5R be constructed having the side $QP_5 = QP_1$ and $RP_5 = RP_2$.

The angle $QP_5R = B$.

It is obvious that $pP_3 = pP_4$; also when the triangle QP_5R is constructed P_5 should be on the line Sp.

PROBLEM II.   Fig. 73.

Given two faces of a trihedral angle and the dihedral angle contained by them, to find the third face and the other dihedral angles.

Let a and β be the two given faces and C their inclination to one another; it is required to find γ, A and B.

Having laid down the two angles a and β as in the last problem, from the point P_1 draw P_1p at right angles to SM, and at the point M make the angle $pMP_2 =$ the given angle C; also make $MP_2 = MP_1$, and from P_2 draw P_2p at right angles to Mp.

Sp is the projection of the third edge of the solid angle; for p is the projection of the point P in the line SP when the two faces are inclined to one another at the angle C.

To find the third face; draw pN at right angles to SN, and from the centre S describe a circle with the radius SP_1, cutting PN produced in the point P_3. NSP_3 is then the rabatment of the angle whose projection is NSp: therefore $NSP_3 = \gamma$.

A and B may now be found as in the last problem.

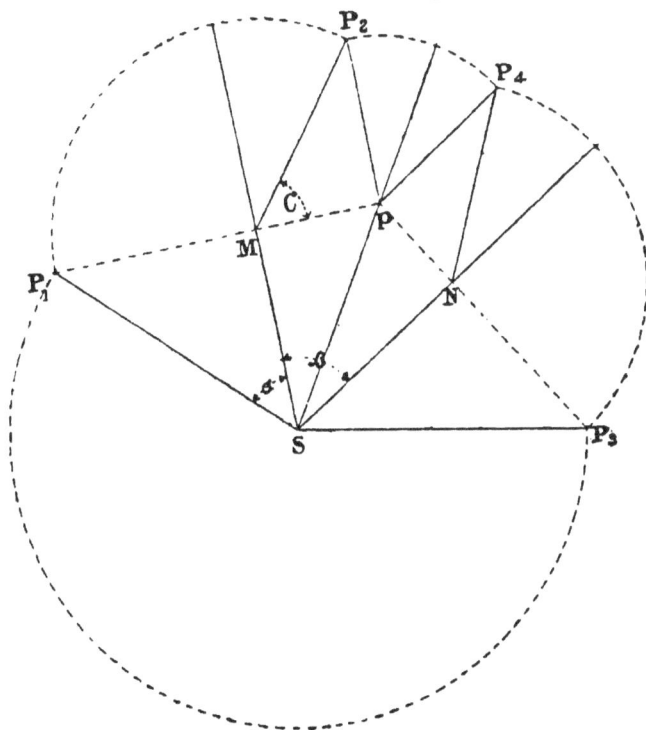

Fig. 73.

PROBLEM III. Fig. 74.

Given two faces of a trihedral angle and the dihedral
angle opposite one of them, to find the third face and the other
two angles.

Let α and β be the two given faces and A the given
dihedral angle.

Having drawn the angles α and β in the horizontal plane,
take a vertical plane at right angles to their common arm
SM; that is, take P_1F at right angles to SM for a ground line.
Now SF being the horizontal trace of a plane, find the vertical trace of the plane when it is inclined at the angle A.
That is, draw eD at right angles to SF, make $eD_1 = eD$, and
the angle $eD_1E = A$. SFE is a plane inclined at the angle
A to the horizontal face β. (See Prob. XIX. Chap. II.)

If the face α revolve about SM till SP_1 lies in the plane
SFE, it is evident a trihedral angle would be formed having
the faces α and β and the angle A.

But as the triangle SMP_1 revolves about SM. P_1 describes
a circle on the vertical plane, since it is at right angles to
SM. Therefore the point P where that circle meets FE is
the vertical trace of the third edge of the trihedral angle.

Sp is the horizontal projection of that edge; and P_2 is
found as in the last problem. $P_2SN = \gamma$.

As the circle P_1PQ cuts FE in two points there are two·
solutions to the problem; that is, two trihedral angles may
be formed from the given data. In the second case Q_1SF is
the third face.

If the circle touched EF there would be but one solution,
and if the line and circle did not intersect the solution would
be impossible—no trihedral angle could be formed from the
given angles. This problem corresponds with what is known
as the ambiguous case in the solution of plane triangles.

Fig. 74.

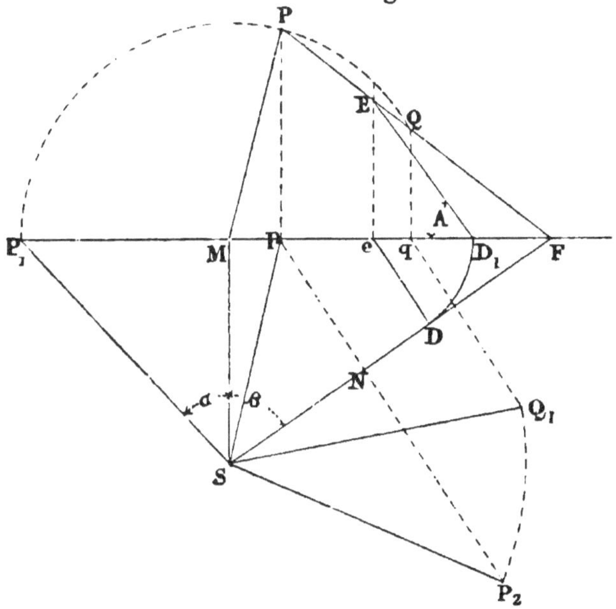

THEOREM. Fig. 75.

If from any point P within the trihedral angle S, perpendiculars Pp, Pq, Pt are drawn to the faces, a trihedral angle is formed at P such that its faces are the supplements of the dihedral angles of S and its dihedral angles the supplements of the faces of S.

Proof. The plane $PpNt$ which contains the perpendiculars Pp and Pt is perpendicular to the two faces which intersect in SN.. (Theor. IV.); and therefore at right angles to SN...(Theor. VI.).

Therefore the dihedral angle between the two faces whose common section is SN is measured by the angle pNt. But in the quadrilateral $PpNt$, the angles at p and t are right angles; therefore pPt and pNt are supplementary.

In the same way it may be proved that the dihedral angles at M and R are the supplements of the angles pPq and tPq, respectively.

Again, as the edges SN, SM, SR are respectively perpendicular to the three faces of the trihedral angles at P, from what has just been proved the dihedral angles of P are the supplements of the faces of S.

The angles S and P are called supplementary.

By means of the principle established in this Theorem the last three cases of the solution of the trihedral angle may be reduced to the first three. When the three angles A, B and C are given, $180 - A$, $180 - B$, $180 - C$ are the faces of the supplementary angle. Having found as in the first case the inclinations of the faces of this supplementary angle which may be called A', B', C'; then $180 - A'$, $180 - B'$, $180 - C'$ are the faces of the trihedral angle whose inclinations are A, B and C.

Similarly, the fifth case resolves itself into the second and the sixth into the third.

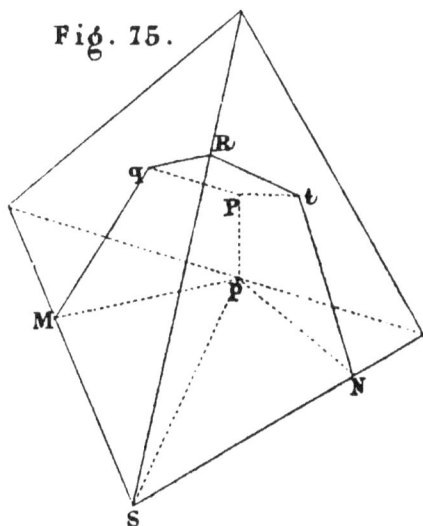

Fig. 75.

Exercises.

1. Draw the plan of a square of $1\frac{1}{2}''$ side in each of the following positions; also an elevation on a plane parallel to one of the diagonals :—

(1) The plane of the square inclined at $50°$ and one of the sides at $30°$.

(2) One side inclined at $40°$ and another at $30°$.

(3) Three consecutive angular points $1''$, $1\frac{1}{2}''$ and $1\frac{3}{4}''$ respectively above the horizontal plane.

2. The side of a regular pentagon $ABCDE$ is $1''$; the heights of the three points A, C, D are $1''$, $1\frac{1}{2}''$ and $2''$ respectively above the horizontal plane : draw its plan, and an elevation on a plane parallel to CD.

3. The three sides of a triangle ABC are, $AB = 2''$, $BC = 2\frac{1}{2}''$, $AC = 3''$; the inclinations of AB and AC to the horizontal are $40°$ and $30°$ respectively: draw the plan of the circle passing through the three points A, B, C, and its elevation on a plane parallel to AB.

4. The plane of a circle of $2''$ radius is inclined $60°$ to the horizontal and $45°$ to the vertical plane of projection ; the centre of the circle being $3''$ from each plane of projection : draw the plan and elevation of the circle.

5. Draw the plan and elevation of a regular pyramid with a square base in the following positions—side of square $2''$, altitude $3''$—

(1) Resting with base on the horizontal plane and one edge of the base making an angle of $30°$ with *xy*.

(2) Resting on one edge of the base, with its axis parallel to the vertical plane of projection and inclined at $60°$ to the horizontal plane.

(3) The axis inclined at 60° to the horizontal and its horizontal projection making an angle of 30° with *xy*.

(4) Resting with one of the triangular faces on the horizontal plane and the horizontal projection of the axis making an angle of 30° with *xy*.

In each of these positions show the projections of a section by a plane which cuts three consecutive edges at distances of 1″, 1½″ and 1¾″ respectively from the vertex. Also show the true form of the section.

6. Find the projection of a cube of 2″ edge,—

(1) On a plane inclined at 40° to one of the faces, and at 30° to a diagonal of that face.

(2) On a plane perpendicular to one of the diagonals of the cube.

7. Draw a plan and elevation of a regular tetrahedron, of 2″ edge, when three angular points are 1″, 1½″ and 2¼″, respectively, above the horizontal plane of projection.

8. One diagonal of an octahedron of 2″ edge is inclined at 30° and an adjacent edge is inclined at 45°: draw its plan and an elevation on a plane not parallel to any edge of the solid.

9. Show the true form of the section of the octahedron in the last exercise by a plane parallel to one of the faces and one inch distant from that face.

10. The side of the base of a regular hexagonal prism is 1½″ and its axis is 3″: draw its plan and elevation when the axis is inclined at 30° to the horizontal plane and 45° to the vertical plane of projection, one edge of the base being in the horizontal plane.

CHAPTER IV.

A SURFACE is generated by the motion of a line. That line is called the *generator* of the surface; it may be straight or curved, and its form and magnitude constant or variable. To determine the surface it is necessary to know the form and magnitude of its generator in every position and the laws which regulate its motion.

The generator may be guided or directed in its motion by lines fixed in magnitude and position, which are called the *directrices;* or it may revolve about a fixed axis. In the latter case the surface generated is called a *surface of revolution.*

This may be illustrated by the case of the plane. It may be generated by a straight line which moves so as to intersect two fixed lines either intersecting or parallel; or which moves along one straight line so as to be always parallel to another fixed line; or it may be generated by a straight line which revolves about a fixed axis at right angles to it.

A surface which can be generated by a straight line is called a *ruled surface.*

Ruled surfaces are divided into two classes, *developable surfaces* and *skew surfaces.*

A developable surface is one which being supposed flexible but inextensible can be unfolded into a plane without tearing or crumpling.

Those surfaces which are the most important from a practical point of view belong either to ruled surfaces or surfaces of revolution.

TANGENT PLANES.

In the curve AB_1B (fig. 76), let any straight line AB be drawn cutting the curve in two points. If that line be con-

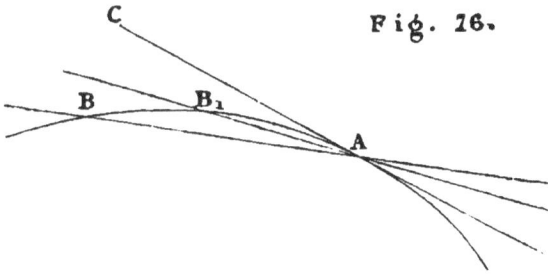

Fig. 16.

ceived to turn about the point A, the other point of intersection will approach A, and there is a certain position which the line approaches as the two points come closer to one another, which it reaches when the two points coincide. The straight line in this position is called a *tangent* to the curve.

If a curve be traced on a surface through any point on it, the tangent to the curve at that point is said to be a tangent to the surface.

THEOREM I.

The tangents to a surface at any point are all in the same plane.

Let A (fig. 77) be a point on a curved surface; BLC, DAE two positions of the generator near to one another;

Fig. 77.

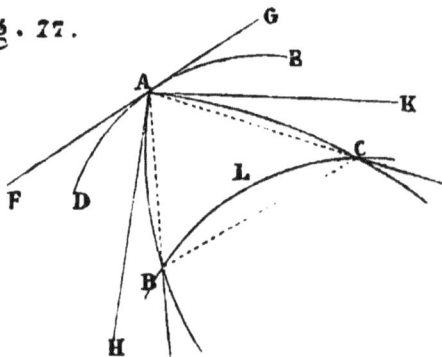

AB, AC any two curves whatever traced on the surface, through A, and meeting the generator BLC at B and C respectively.

The three straight lines AB, AC, BC are in the same plane however near BLC may be to DAE, and must therefore be in the same plane in the limit when BLC coincides with DAE. But when BLC coincides with DE, the points B and C coincide with A, and the three straight lines become the tangents AH, AF and AK, to the three curves through A. Therefore these three tangents are in the same plane.

In a similar manner it may be shown that any other curve on the surface passing through A has its tangent in the plane of AF, AH.

The plane which contains all the tangents to a surface at any point is called a *tangent plane* to the surface at that point. It is determined by finding the plane containing any two of these tangents.

In a ruled surface one of the lines which can be traced on it is straight, and therefore coincides with its tangent.

Hence a tangent plane to a ruled surface at any point contains the straight generator passing through that point.

The *normal* to a surface at any point is the line perpendicular to the tangent plane at that point.

The outline or boundary of the projection of a surface is the projection of the line containing all the points of contact of the tangent planes to the surface which are at right angles to the plane of projection.

The Cone.

A conical surface is generated by a straight line which passes through a fixed point and moves along a fixed line.

The fixed point of the generator is called the vertex of the cone and the fixed line the directrix.

It is evident that any line traced on a given conical surface might serve as its directrix. Following the definition a cone would be determined by the projections of its vertex and directrix; but for convenience the horizontal trace of the surface will be taken for the directrix. If any other directrix were given the horizontal trace of the surface could be found by determining the horizontal traces of the generator in different positions. The horizontal trace will be called the base of the cone.

When the generator is unlimited in length part of the surface lies on each side of the vertex. These are called the two sheets of the conical surface.

The Cone of revolution, called also a *right circular cone*, is generated by a straight line which revolves about a fixed axis that it intersects, and with which it makes a constant angle. A right-angled triangle revolving about one of its sides fulfils these conditions, and is the common way of defining this surface.

Every point of the generator describes a circle having its plane perpendicular to the axis, which passes through its centre.

PROBLEM I. Fig. 78.

Given one projection of a point on a cone, to determine th other projection.

Let vv' be the projections of the vertex of the cone $ABCD$ its horizontal trace, and p the projection of a poin on the surface; it is required to find the other projection of P.

Construction. Draw vp, the horizontal projection of the generator through P, meeting the horizontal trace of the cone at F, and draw $f'v'$ the vertical projection of the generator VF. p' on the line $f'v'$ is the vertical projection required.

When the horizontal trace of the surface is a closed curve as in the figure there are two solutions; for the line vp meets $ABCD$ in a second point E, so that the projection of the two generators VE and VF pass through p. Therefore the vertical line through p meets the surface in two points.

THEOREM II. Fig. 79.

The tangent plane to a cone at any point is a tangen at every point of the generator passing through that point.

Let B and C be any two points on the same generator it is required to show that the tangent plane to the surface at B is also a tangent at C.

Proof. Let BD and CE be two curves traced on the surface and meeting a generator VD in the points D and E.

Since VB and VD are in the same plane, so also are BD and CE, the secants of the arcs; and as this is true however near VD be taken to VB, it must be true in the limit when the two generators coincide. That is, the tangent BF and CG to the two curves are in the same plane.

But the plane containing VB and BF is the tangent to the surface at B; and the plane containing VB and CG is the tangent plane at C.

Therefore the plane which is a tangent at B is also tangent at C.

Fig. 78.

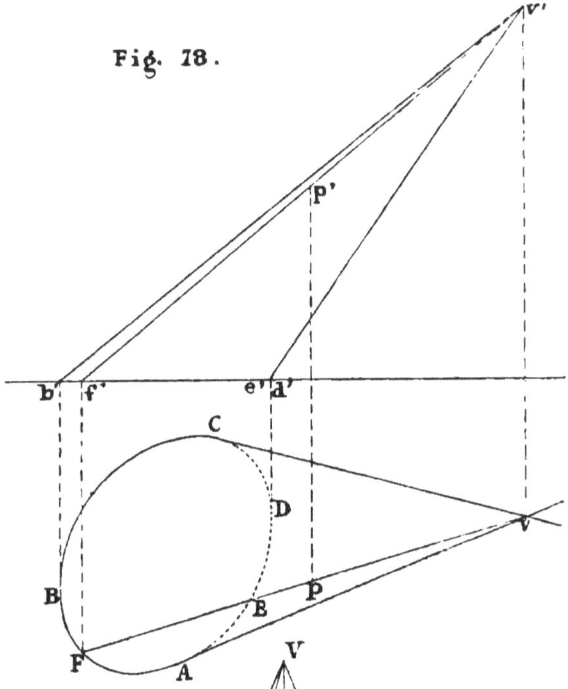

Fig. 79.

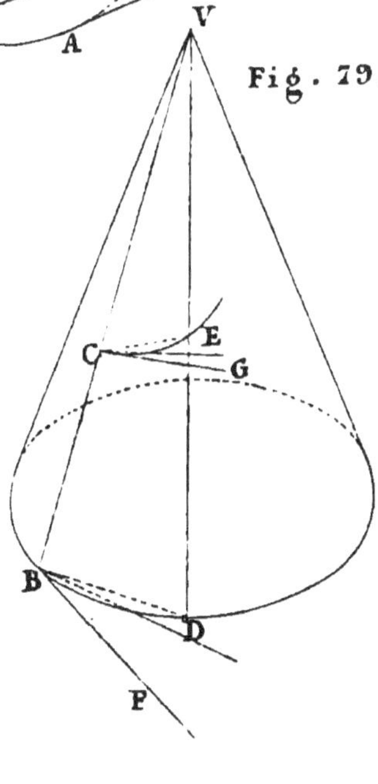

PROBLEM II. Fig. 80.

To draw a tangent plane to a cone through a given point in the surface.

Let *vv'* be the projections of the vertex of the cone, *ABC* its horizontal trace, and *pp'* the projections of a point on the surface ; it is required to find the traces of the tangent plane at the point *P*.

Construction. Draw *vC* and *v'c'* the projections of the generator through *P*. Draw *CM* a tangent to the curve *ABC*; this is the horizontal trace of the plane required.

Next determine the vertical trace of the plane containing *CM* and *CV*—which may be done either by finding the vertical trace of *CV*, or drawing *VN* parallel to *CM*, as shown in the figure. *CMN* is the plane required.

Proof. Since the plane *CMN* contains the tangent to the curve *ABC* at the point *C*, and the generator *CV*, it is a tangent plane to the cone; and it contains *P*, which is in the line *CV*. It is therefore the tangent plane required.

Fig. 80.

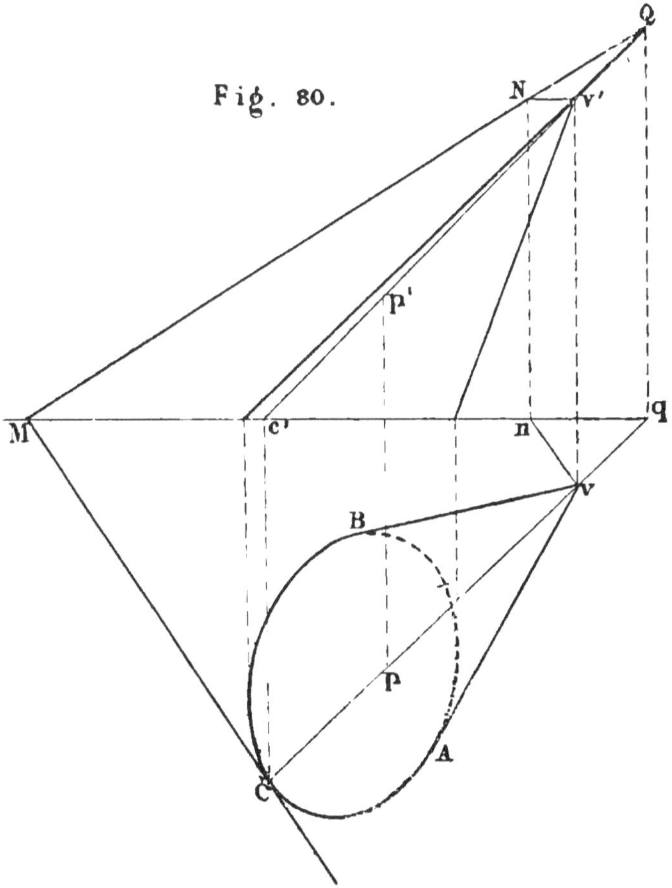

PROBLEM III.   Fig. 81.

To draw a tangent plane to a cone through a given external point.

Let *vv'* be the projections of the vertex, *ABC* the horizontal trace of the cone, and *pp'* the given point; it is required to draw through *P* a tangent plane to the cone.

Construction. Find the traces *L* and *N* of the line *VP*; draw *LM* a tangent to *ABC*, and join *MN*. *LMN* is the plane required.

Proof. Since the plane contains the line *LN* it passes through the points *P* and *V* which are on that line; and as it contains *LM*, the tangent to *ABC* at *D*, and therefore the line *VD*, it is a tangent plane to the cone.

Remarks. Should the vertical trace of *PV* be too remote, a point in the vertical trace of the plane may be found as in the last problem by drawing a line through *V* parallel to *LM*.

As a second tangent to *ABC* may be drawn from *L*, this problem admits of two solutions.

Fig. 81.

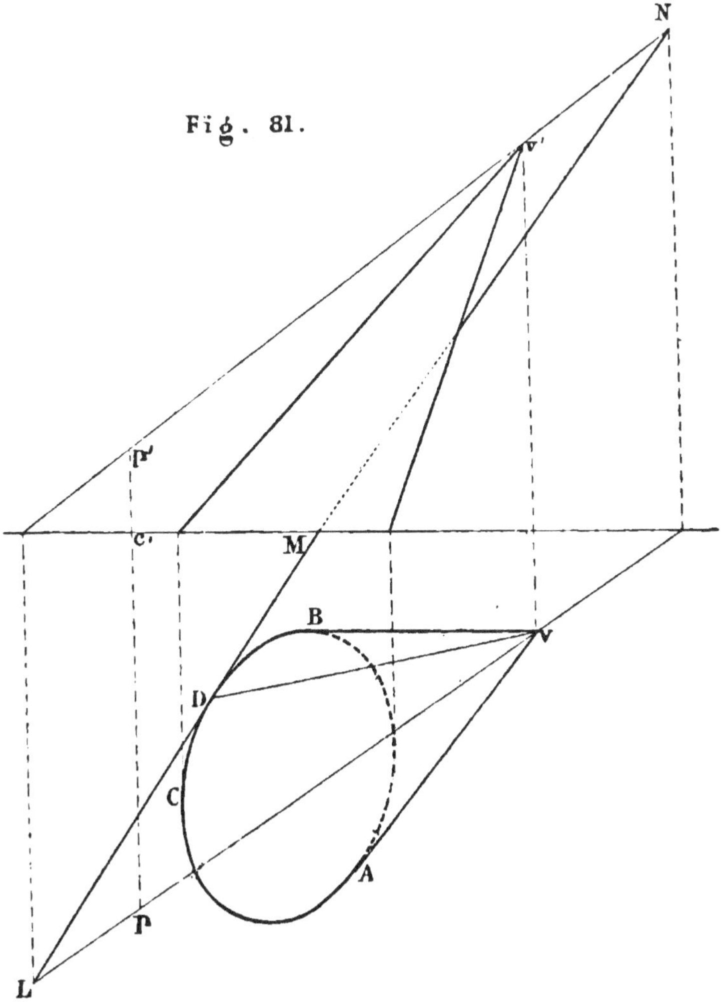

PROBLEM IV.   Fig. 82.

To draw a tangent plane to a cone which shall be parallel to a given straight line.

It is required to draw a tangent plane to the cone *VABC* which shall be parallel to *PQ*.

Construction. Find the traces of a line through *V* and parallel to *PQ*. These are *L* and *N*. Draw *LM* touching *ABC* and join *MN*. *LMN* is the plane required.

Proof. Since *LMN* contains *VN*, a line parallel to *PQ*, *LMN* and *PQ* are parallel to one another (Theor. x. Ch. 1.), and as *LMN* contains the line *VD* and the tangent to *ABC* at the point *D* it is a tangent plane to the cone.

Remark. It may be observed that when a tangent plane touches a surface along a line it can only be made to fulfil one other independent condition, such as passing through a given point, or being parallel to a given line. Thus a plane cannot generally be drawn to touch a cone and to contain a given line, for that would be equivalent to drawing a plane to contain two given lines, which is impossible except the lines are either intersecting or parallel.

Fig. 82.

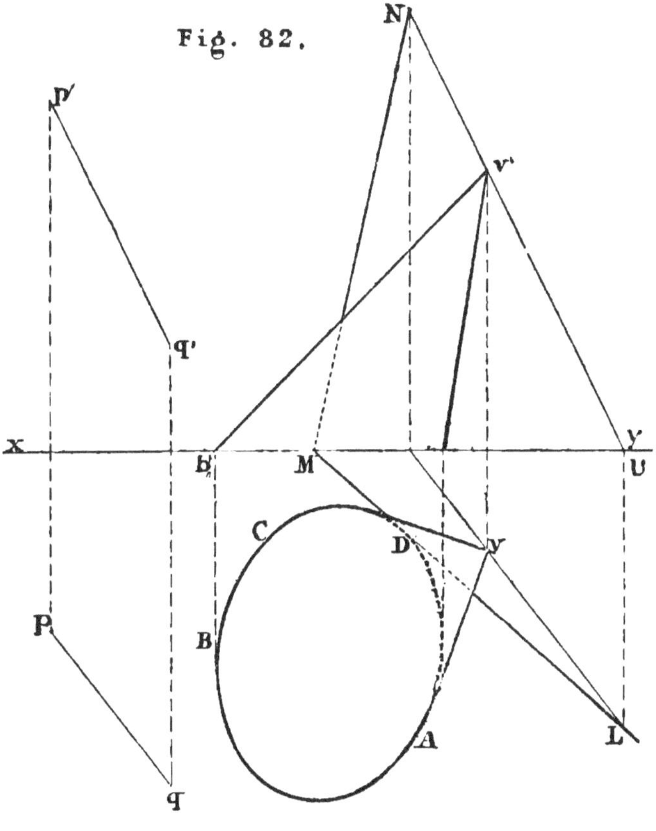

PROBLEM V.   Fig. 83.

To find the traces of a plane which shall contain a given line and have a given inclination.

Let AB be the given line and θ the given angle; it is required to draw the traces of a plane containing AB and having an inclination θ to the horizontal plane.

Construction. Take vv' the projections of any point on AB. Draw $v'c'$ making the angle θ with xy. Now suppose VC to revolve about Vv, generating a right circular cone the trace of which is the circle CD. Draw that circle, and from B, the horizontal trace of AB, draw BE a tangent to it at the point D. BEA is the plane required.

Proof. The plane BEA contains the line AB, since it contains the traces A and B. As it contains the vertex V of the cone, its inclination is the angle between VD and vD. (Prob. XIX. Ch. II.)

But the angle $VDv = \theta$ by construction.

Therefore BEA is the plane required.

As two tangents to the circle may be drawn from B there are two solutions to this problem.

Corollary. The angle between a tangent plane to a cone of revolution and a plane at right angles to its axis is equal to the angle contained by the hypotenuse and base of the right-angled triangle which generates the cone.

Fig. 83.

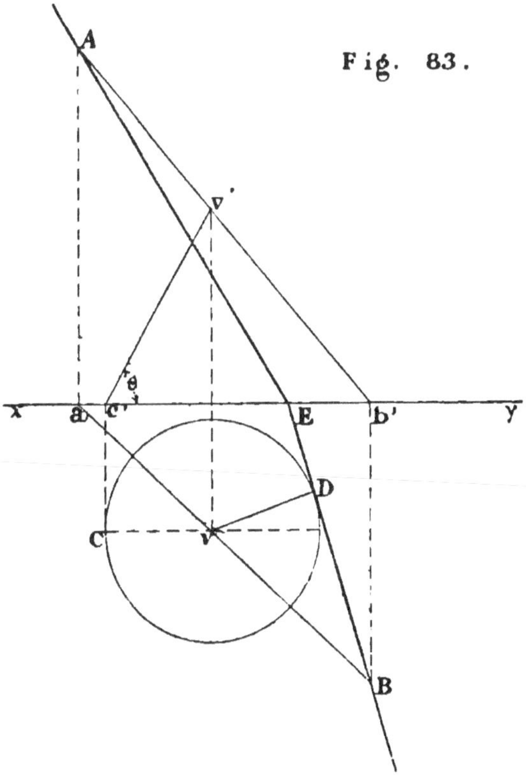

PROBLEM VI. Fig. 84.

To find the traces of a plane which shall contain a given line and make a given angle with a given plane.

Let LMN be the given plane and AB the given line; it is required to find the traces of a plane containing AB and making an angle θ with LMN.

The method of solving this problem is suggested by the corollary to the preceding one. For if a cone of revolution be conceived, having its vertex in the given line, its axis perpendicular to the given plane, and its generators making an angle θ with that plane, then the tangent plane to the cone which should also contain the given line would be the plane required. Hence the following :—

Construction. Through any point A of the given line draw a vertical plane DEL at right angles to the given plane. Let DEL be turned into the vertical plane of projection, bringing with it the point A and the line of intersection DL, these are now A_1 and D_1L. From A_1 draw A_1G_1 at right angles to D_1L, and A_1H_1 making an angle θ with D_1L. G_1H_1 is the radius of the circular trace of the cone with LMN. Now the line LD_1 being returned to its original position, and the plane LMN rabatted on the horizontal, the point G comes to G_2, and H to H_2. This rabatment is most easily effected by producing ED and making DG_2 equal to D_1G_1, and DL_1 equal to D_1L. With centre G_2 and radius G_2H_2 describe the circle H_2KO, which is the rabatment of the trace of the cone with LMN. Next find the point of intersection of AB with LMN and its rabatment P_1, and draw P_1Q a tangent to the circle. NQ is the rabatment of the line of intersection of the required plane with LMN.

NRQ_1, which contains the two intersecting lines AB and NQ_1, is the plane required. The second tangent to the circle gives a second plane whose traces are STU.

If the given line and plane were parallel to one another, the line of intersection of the tangent plane to the cone with LMN, that is the line NQ_1 in the figure, would be parallel to the given line. (Theor. XI. Ch. I.)

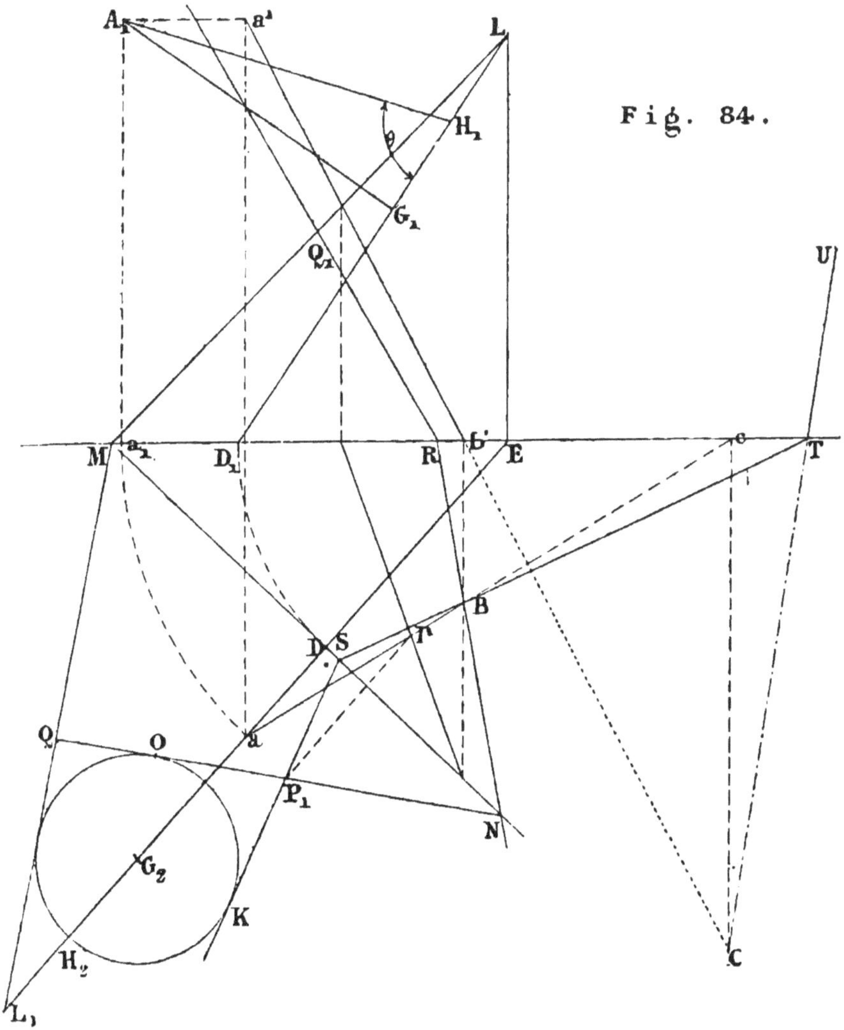

Fig. 84.

PROBLEM VII. Figs. 85, 86 and 87.

To find the section of a cone of revolution by a plane.

First case, Fig. 85. Let ABC, $a'b'c'$ be the projections of a cone of revolution, and LMN the traces of a plane *which cuts all the generators*, but not at right angles to the axis; it is required to find (1) the projections of the curve of intersection of the plane with the cone, and (2) the true form of that section.

Construction. The horizontal plane is taken at right angles to the axis of the cone, and the vertical plane at right angles to LMN. The vertical projection of the required curve is therefore the straight line $o'N$.

To find the horizontal projection of any point P of the curve, when the vertical projection p' is known; draw, as in Prob. I., the vertical projection $v'p'$ of the generator through P, and determine the horizontal projection Dv of that generator; the point p on Dv is the horizontal projection required.

In the same way any number of points on the horizontal projection of the curve may be found and the curve traced through them.

This method fails for the point q which is best determined by drawing the horizontal projection of the circle described by the point Q as it revolves about the axis; $q'r'$ is the radius of that circle; therefore the point q is on the circle described about v with the radius vr.

Next, to find the true form of the section, the plane LMN is rabatted on the horizontal plane of projection, which determines the curve $N_1Q_1P_1O_1S_1$. That curve is an *ellipse*.

The curve opn is also in general an ellipse, though it may be a circle.

Fig. 85.

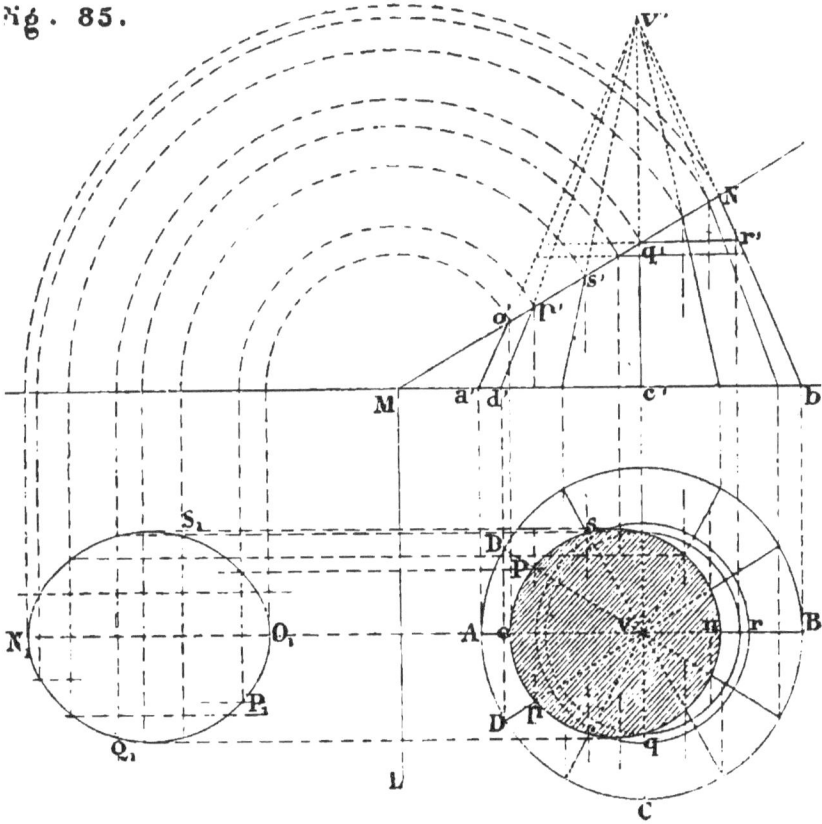

Second case, Fig. 86. Let the cutting plane *LMN* be parallel to one generator, and one only. In the figure it is parallel to *A V*.

The method of determining the curve is the same as in the first case, and is evident from the figure. The curve *ON,L* is a *parabola—OnL* is also a parabola.

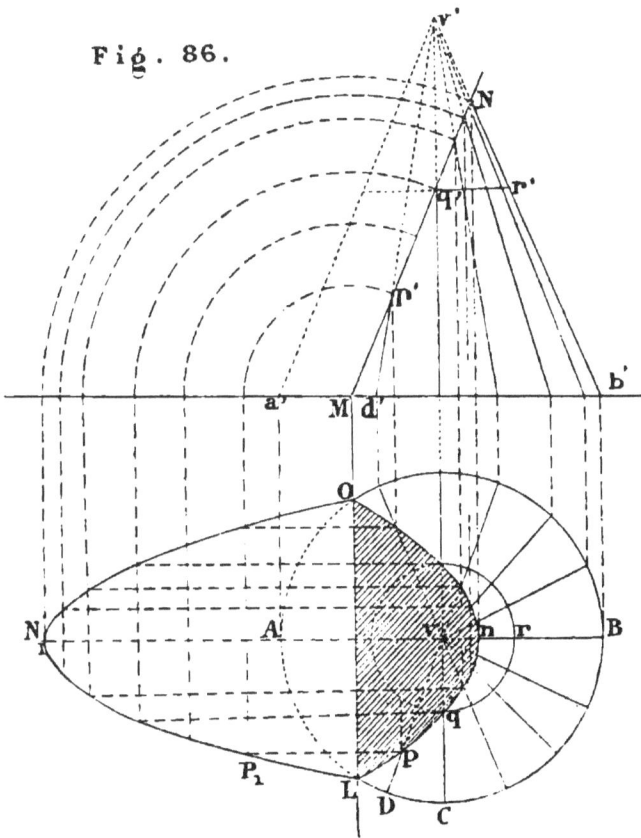

Fig. 86.

Third case, Fig. 87. Let the cutting plane *LMN* be parallel to two generators *DE* and *FG*, or, what comes to the same thing, let it cut both sheets of the conical surface. The curve of intersection has then two branches. That curve is the *hyperbola*.

The horizontal projection of the curve might be found as in the other two cases, but as the straight generators cut *LMN* so obliquely a better solution is obtained by taking a series of horizontal circles on the cone. The rest of the construction will be evident from the figure. The curve is for convenience rabatted on the vertical plane.

The *tangent* to the plane section of a curved surface at any point is the line of intersection of the cutting plane and the tangent plane at that point. The tangent to any of these conic sections at a given point may be readily obtained in that way.

The *asymptotes* of the hyperbola, that is to say the tangents at points infinitely distant, when the conical surface is extended indefinitely, are the lines of intersection of *LMN* with the tangent planes along *DE* and *FG*; *LP*, *QR* are the projections of the asymptotes, and L_1O_1, O_1Q_1 their rabatments.

The general problem of finding the section of any conical surface by a given plane is solved by finding the points of intersection of a sufficient number of straight lines, which are the generators, with the given plane, by Prob. XIV. Chap. II.

The ellipse diminishes in size as the cutting plane approaches the vertex, and its limit is a point. The limit of the hyperbola is a pair of straight lines intersecting at the vertex; the limit of the parabola is a straight line.

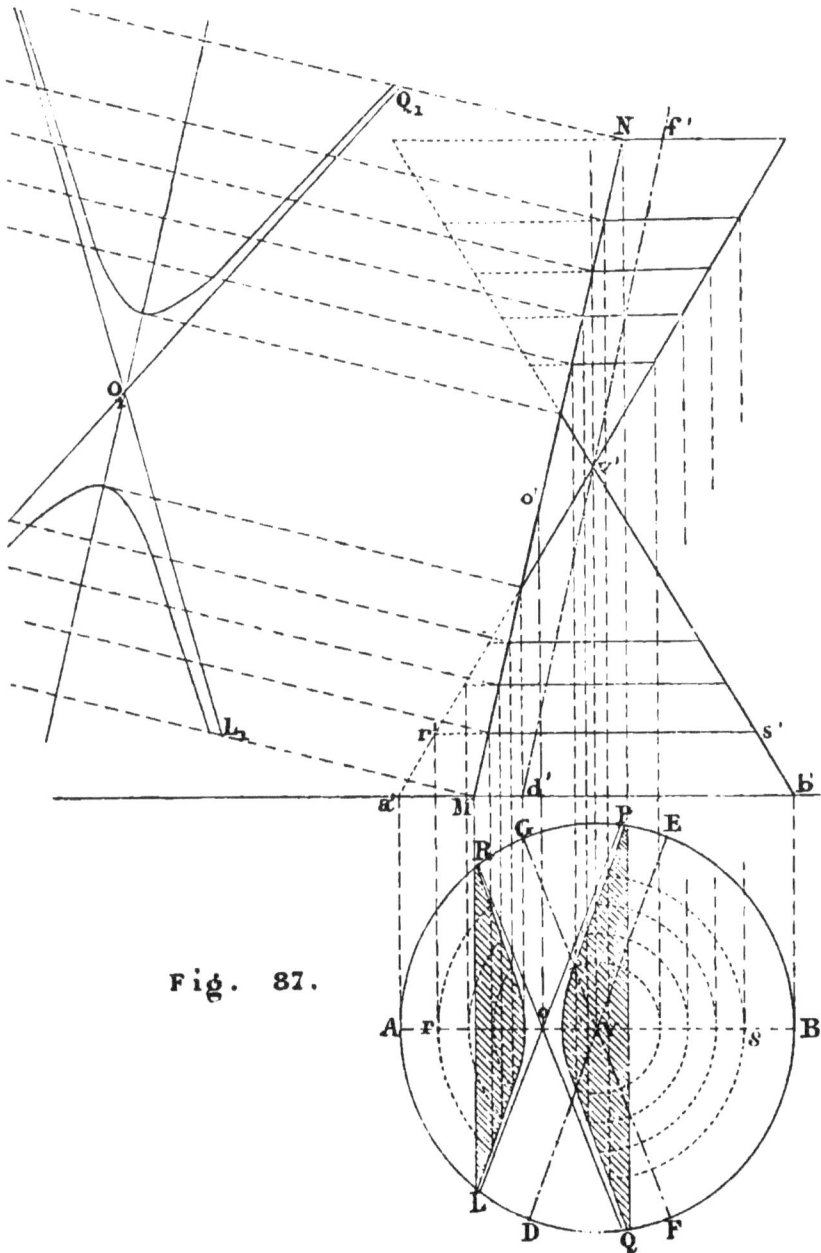

Fig. 87.

PROBLEM VIII. Fig. 88.

To find the development of a given conical surface.

As a curved line may be considered as the limit of a polygon, when the number of its sides is increased indefinitely, so a cone may be considered as the limit of a pyramid when the number of its faces is increased indefinitely.

Let *VABCDE*, fig. 89, be a conical surface, and suppose the triangle *VAB* to turn about *VB* till *VAB* and *VBC* are in the same plane, and then these two faces together to turn about *VD* till the three triangles are in the same plane. By proceeding in this way and the points *A*, *B*, &c. being taken sufficiently near one another, the polygon obtained when all the triangles are brought into the same plane may be made to approach indefinitely near to the true development of the curved surface.

In a right circular cone the development is a sector of a circle, for the lines *VA*, *VB*, &c. are all equal to one another.

Fig. 88 is the development of the cone in fig. 85, and the curve *OPQN* is the development of the ellipse. The cone is supposed to be divided along the line *AV*, and the points *O*, *P*, *Q* are found by setting off on *VA*, *VD*, *VC* of the development the distances of the same points from the vertex of the cone.

THE CYLINDER.

A cylindrical surface is generated by a straight line which moves so as to be always parallel to a fixed line.

A cylinder is completely defined when the direction of one of its generators and the directrix are given.

A *cylinder of revolution* is generated by a straight line revolving about an axis parallel to it. Every point of the

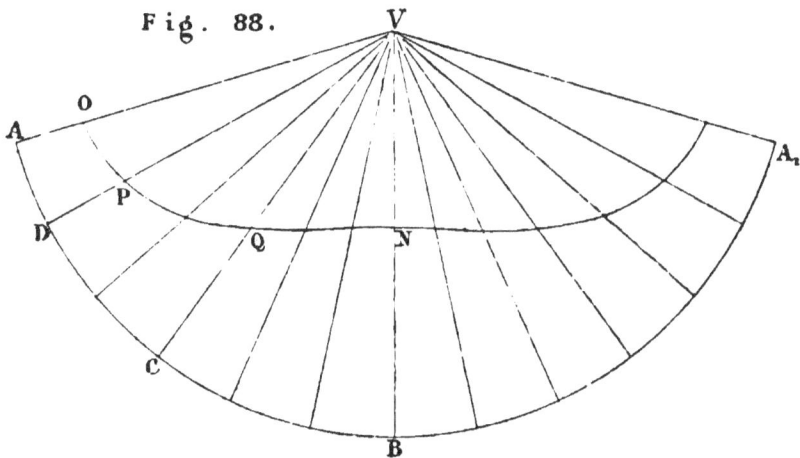

Fig. 88.

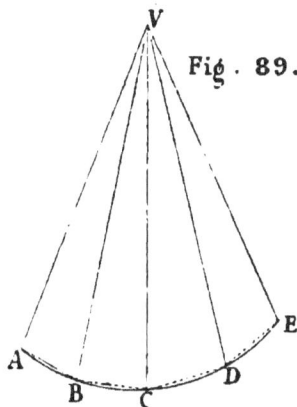

Fig. 89.

revolving line describes a circle whose plane is at right angles to the axis, hence it is also called *a right circular cylinder.*

A cylinder may be regarded as the limit of a cone of which the vertex is at an infinite distance, for the generators of the cone are then parallel to one another. Therefore all the properties of the cone which are independent of the position of the vertex are also properties of the cylinder. Hence the following may be at once affirmed :—

1. The tangent plane to a cylindrical surface at any point is a tangent along the whole length of the generator passing through that point[1].

2. The section of a cylinder of revolution by a plane which cuts the axis obliquely is an ellipse.

When a plane cuts a cylindrical surface along one or more generators it is parallel to all the others (Theor. X. Chap. I.), and conversely if it be parallel to the generators, the section will be one or more straight lines.

The section of a cylinder by a plane at right angles to the generators is called a right section.

It has not been thought necessary to work the problems for the cylinder corresponding to I., II., III. and IV. for the cone, where a trace of the surface is known, for they would almost be a repetition of those which have been already given; but the three following problems, which are important special cases, have been substituted.

---

[1] It follows from this property of a cylinder, that if a line be a tangent to a curve, the projection of the line is a tangent to the projection of the curve. For the projection of the curve is the trace of its projecting cylinder, and the projection of the tangent is the trace of the tangent-plane to the cylinder.

PROBLEM IX.   Fig. 90.

Given one projection of a point on a cylinder of revolution, to find the other projection, when the axis of the cylinder is parallel to the ground line.

Let *abcd*, *e'f'g'h'* be the projections of a cylinder of revolution, and *p* the horizontal projection of a point on the surface; it is required to find the vertical projection of the point *P*.

Suppose the cylinder to be cut by a plane at right angles to the axis and containing the point *P*. The section is a circle, the projections of which are the two straight lines *mn* and *o'q'* at right angles to the projections of the axis. If that circle be turned about its horizontal diameter *MN* till it is parallel to the horizontal plane, the point *P*, moving with the circle, has p_1, p_1' for its projections. If now the circle be returned to its original position, p_1' comes to *p'* which is the vertical projection of the point *P* on the cylinder. Hence the following :—

Construction. Draw *mq'* at right angles to the ground line. On *mn* as diameter describe a circle; draw through *p* a line parallel to *xy* meeting the circle in p_1, and make *m'p'* equal to pp_1.

p' is the vertical projection required.

There are evidently two points on the cylinder which have *p* for their horizontal projection.

Fig. 90.

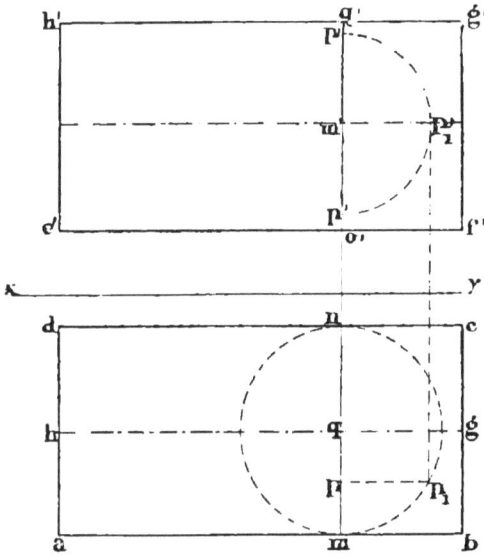

PROBLEM X. Fig. 91.

*Through a given point on a cylinder of revolution to draw
a tangent plane to the surface, when the axis of the cylinder
is parallel to the ground line.*

Let *abcd*, *e′f′g′h′* be the projections of a right circular
cylinder, and *p* the horizontal projection of a point on the
surface; it is required to draw a tangent plane at the
point *P*.

Construction. Draw LR_1, the traces of a plane cutting
the cylinder at right angles to the axis. Let that plane be
rabatted on the horizontal plane, and find the position of *O*,
the centre of the circular section of the cylinder, and of *P*,
the given point on the surface, in the rabatment of the
plane...... (Prob. XXIV. Chapter II.)

These are the points O_1 and P_1 respectively.

Describe the circle O_1P_1; draw the tangent P_1L, meeting
xy in *R*; and make $KR_1 = KR$.

LM and R_1N, parallel to *xy*, are the traces of the plane
required.

Proof. The line *LR* is the rabatment of the tangent to
the circular section of the cylinder *OP*, by construction;
that is to say, the line whose traces are *L* and R_1 is a tangent
line to the surface at the point *P*.

But the plane *LM* R_1N contains the line LR_1, and
therefore passes through *P*; and passing through *P* it
contains the line through that point parallel to its traces,
which is the generator *PS*.

Therefore *LM* R_1N is a tangent plane to the cylinder at
the point *P*.

PROBLEM XI. Fig. 91.

*Through a given external point to draw a tangent plane to
a cylinder of revolution, when the axis is parallel to the ground
line.*

Let *abcd*, *e′f′g′h′* be the projections of the cylinder and
q, *q′* the projections of the point; it is required to draw
through *Q* a tangent plane to the cylinder.

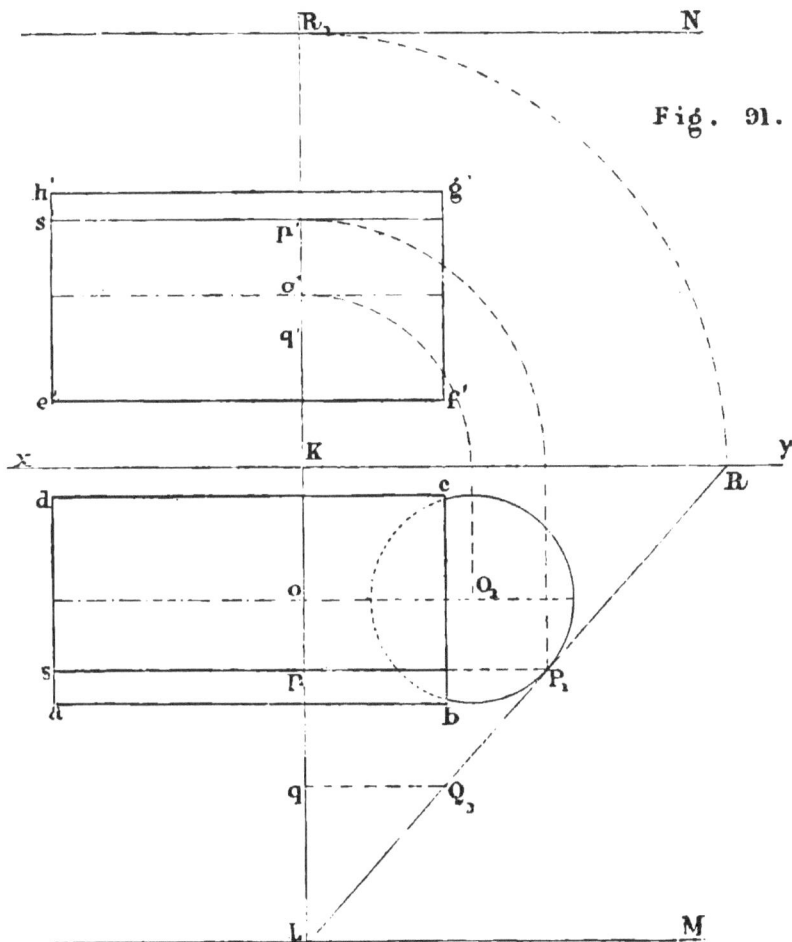

Fig. 91.

138   DESCRIPTIVE GEOMETRY.

Construction. Draw, as in Prob. x., LR_1 the traces of a plane containing Q and perpendicular to the axis of the given cylinder. Rabat that plane on the horizontal, and determine O_1 and Q_1, the rabatments of O and Q. With O_1 as centre describe a circle equal to the circular section of the cylinder, and draw Q_1R a tangent to the circle. The remaining part of the solution is the same as in the preceding problem.

Problem XII.   Fig. 92.

To find the development of a right circular cylinder.

Let ABC, $a'c'f'd'$ be the projections of a right circular cylinder; it is required to find its development.

A cylinder may be considered as a prism the faces of which are indefinitely small, that is to say, it is the limit towards which the prism approaches as the number of its faces is indefinitely increased. When the cylinder is right, as in this case, the generators being all at right angles to the circular base, each of the small faces is a rectangle. Hence the following :—

Construction. Divide the circumference of the circle ABC into a number of equal parts $C\ 1, 12$, &c., and set off on a straight line $c'C_1$ distances equal to $C\ 1, 12$, &c., or in any other way make $c'C_1$ equal to the circumference ABC. The rectangle $c'C_1F_1f'$ is the development required.

Cor. To find the position of any point P on the development, when its projections $4p'$ on the cylinder are given. Find the development of the generator passing through the point, and make $4P_1 = 4'p'$. In this way the development of any curve on the surface may be found.

Any curve on a right cylinder such that its development is a straight line oblique to the generators is called a *helix*. $c'g'f'$ is the projection of such a curve, $c'F_1$ being its development. It manifestly cuts the generators of the cylinder at equal angles, and is the well-known form of a screw-

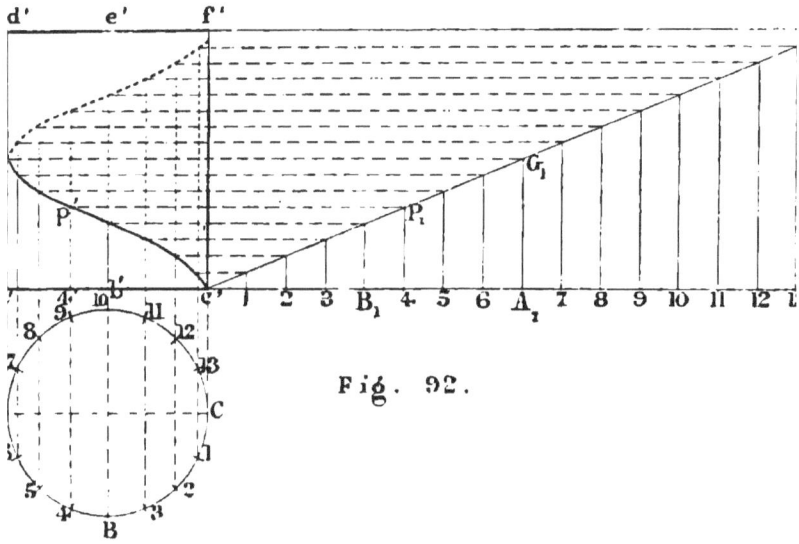

Fig. 92.

thread. A *screw surface* is generated by a straight line moving on the axis of a cylinder and a helix traced on its surface, the generator making a constant angle with the axis. The surface of a triangular screw is formed when the generator cuts the axis obliquely, and the square threaded screw when the generator cuts the axis at right angles. The latter surface is used for the coursing joints of oblique arches.

SURFACES OF REVOLUTION.

In figure 93, let XY be a fixed straight line, and ABC any curve whatever. Let perpendiculars AO, BP, CQ be drawn from the different points of ABC to XY. Now if the whole system of lines ABC OPQ turn about the axis XY, each of the points A, B, C describes a circle having its centre in the axis.

The surface generated by ABC is called a *surface of revolution*.

The following properties of a surface of revolution are manifest from the way in which it is generated.

1. The lines AO, BP, &c. move in planes at right angles to XY; hence any section of a surface of revolution by a plane at right angles to the axis is a circle. These circles are called *parallels*.

2. All sections of the surface by planes containing the axis, as $A_1B_1C_1$, are equal and similar curves. They are called *meridians*, and a plane containing the axis is called a *meridian plane*. It follows that any surface of revolution may be generated by a meridian revolving about the axis. A surface of revolution will be considered as given when its axis and a meridian are given.

3. At any point B_1 of the surface the tangent B_1T to the parallel, being perpendicular to the axis (Theor. III. Ch. I.) and to the radius PB_1, is at right angles to the plane containing these two lines, which is the meridian plane at the point B_1. The tangent plane at B_1, which contains B_1T must also be perpendicular to the meridian plane....(Theor. IV. Chap. I.) Therefore, *the tangent plane to a surface of revo-*

lution is perpendicular to the meridian plane which passes through the point of contact.

4. The normal to the surface at any point, being perpendicular to the tangent plane, lies wholly in the meridian plane (Theor. v. Cor. Ch. I.) and consequently either intersects the axis or is parallel to it.

5. If the meridian plane PB_1R turns about the axis XY, the straight line B_1R will generate a cone of revolution, and the generator B_1R will be in every position a normal to the surface. Thus, *the normals to a surface of revolution at the different points of the same parallel form a cone of revolution having its vertex on the axis.*

6. When the tangent to a meridian intersects the axis it generates a cone of revolution as the meridian moves about the axis. Therefore, *the tangents to a surface of revolution at the different points of the same parallel meet the axis in the same point.* When the tangents are parallel to the axis a cylinder of revolution is generated.

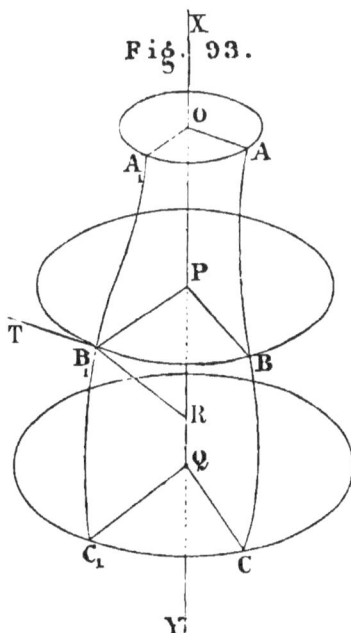

Fig. 93.

PROBLEM XIII. Fig. 94.

Given one projection of a point on a surface of revolution, to determine the other projection.

It is convenient to take one of the planes of projection perpendicular to the axis. Let this be the horizontal plane, then the *parallels* are all projected into circles on the horizontal plane, and into straight lines on the vertical plane.

Let p be the horizontal projection of a point on the surface; it is required to find the vertical projection of the point.

Construction. With centre a describe the arc pp_1. This is the horizontal projection of the parallel on which the point lies. Next draw p_1p_1' cutting the meridian $a'b'c'$ in the point p_1'. $p_1'q$ is the vertical projection of the parallel through P. Therefore p' is the vertical projection required. In this figure there are clearly two solutions, for the point p'' may also be the vertical projection of p.

PROBLEM XIV. Fig. 94.

To draw a tangent plane to a surface of revolution at a given point on the surface.

Let p, p' be the projections of a point P on the surface; it is required to draw a tangent plane to the surface at P.

Construction. Determine, as in the last problem, the projections of the parallel containing the given point. At the point p_1' draw the tangent $p_1'v'$ and normal $p_1'o'$ to the meridian $a'b'c'$. As the normal to the surface at every point of the parallel through P passes through the point O in the axis (5), $o'p'$, ap are the projections of the normal at P.

Draw the plane LMN through P, and at right angles to the normal at that point......(Prob. XVI. Ch. II.). LMN is obviously the tangent plane required.

It may be observed that the tangent plane to the given

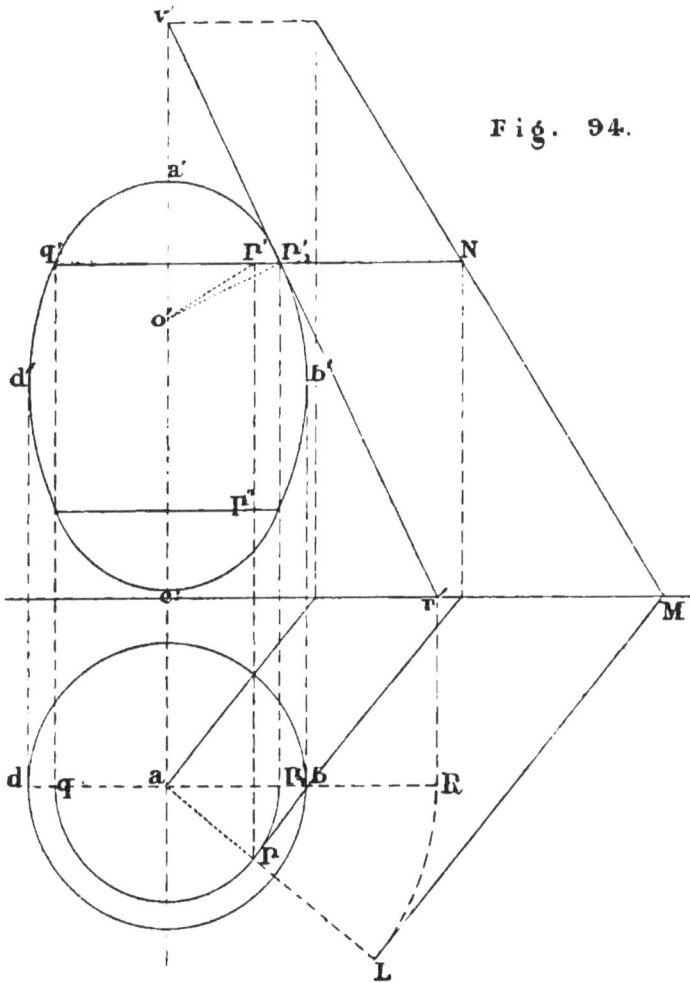

Fig. 94.

surface at P is also the tangent plane at the same point to the cone generated by the revolution of PV about the axis OV. In fact the two surfaces touch each other along the parallel QP, that is to say, they have a common tangent plane at every point of that line.

From this it is clear that the plane LMN might be obtained by drawing a tangent plane to the cone of revolution with vertex V, axis VA, and generator VR, by Problem II.

The construction for this solution is also shown in the figure.

THE SPHERE.

If a circle revolve about one of its diameters as an axis which is fixed in position during the motion, the surface generated is a sphere.

Every point of a spherical surface is at the same distance from the centre of its generating circle, for it is a point on that circle in one of its positions.

Hence a spherical surface may be defined as one of which every point is at the same distance from a point within it. That point is called the centre of the sphere.

THEOREM III.

Every plane section of a sphere is a circle.

Proof. If from the centre of the sphere a perpendicular be drawn to the cutting plane, and a line to any point of the section, these two lines will form the two sides of a right-angled triangle, the third side of which is in the cutting plane; and as the two former sides are both constant for every point of the curve, the latter must be constant also: that is to say, every point of the line of section is at the same distance from the foot of the perpendicular from the centre to the cutting plane. The section is therefore a circle, the centre of which is the foot of the perpendicular from the centre to the cutting plane.

The section of a sphere by a plane through the centre is called a *great circle*.

The diameter of a great circle of a sphere, that is any line through the centre of the sphere and terminated by the surface, is called a diameter of the sphere.

Any great circle of a sphere may be considered as its generator, and any diameter of that circle as the axis. Hence from the property of surfaces of revolution (6) the tangents to a sphere through an external point form a cone of revolution which touches the sphere along a circle.

The line joining any point on the surface of a sphere with the centre is a normal to the surface, for it is at right angles to the tangents of all great circles which meet it.

PROBLEM XV. Fig. 95.

Given the projections of four points on the surface of a sphere, to find its centre and radius.

Let the four points be *A, B, C,* and *D.*

Construction. Draw the projections of the three lines *AB, AC, AD*; and determine the traces of the three planes *LMN, LRS, TUN,* each of which passes through the middle point of one of the lines and is perpendicular to it ......
(Prob. XVI. chap. II.)

(To prevent confusion in the figure the construction lines for these planes have been omitted.)

Thus *LMN* is perpendicular to *AB,* and passes through its middle point *E*; *TUN* bisects *AC,* and is at right angles to it; and *LRS* bisects *AD* and is at right angles to it. *O,* the point of intersection of these three planes, is the centre of the sphere.

The radius is equal to the distance between *O* and any of the given points.

Proof. Since $AE = EB$, and *O* is in the plane *LMN,* the two right-angled triangles *OEA, OEB* are equal in all respects; therefore $OA = OB$.

Similarly it may be shown that $OC = OA = OD$.

Therefore the four points *A, B, C, D* are on the surface of a sphere of which *O* is the centre.

PROBLEM XVI.

To draw a tangent plane to a sphere at a given point on the surface.

Construction. Draw a plane through the given point at right angles to the radius passing through it. That is the tangent plane required.

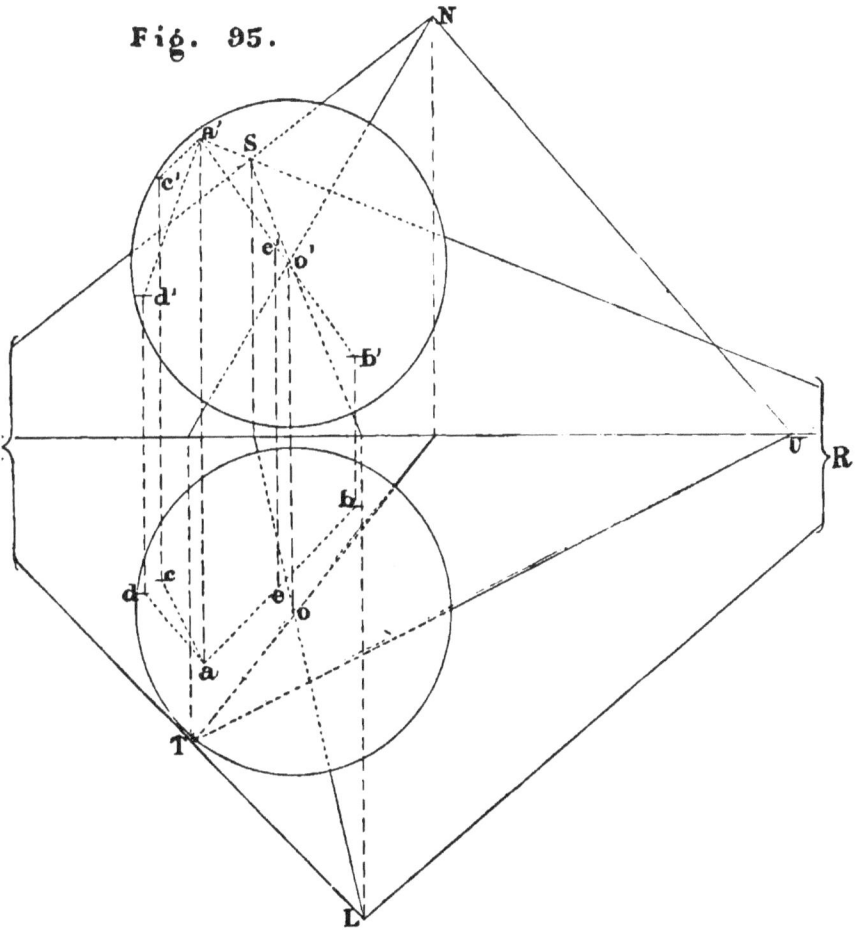

Fig. 95.

PROBLEM XVII.   Fig. 96.

*Given the projections of a sphere and of an external point,
to find the projections of the circle of contact of the cone which
envelops the sphere and has the given point for its vertex.*

Let oo' be the projections of the centre of the sphere, of
which the radius is equal to oa; and vv' the projections of the
point; it is required to find the projections of the circle of
contact of all the lines which can be drawn through V to
touch the given sphere.

If any plane whatever be drawn containing V and cutting
the sphere, and if the point of contact of a tangent through
V to the circle of section of the sphere by the plane be
determined, that will be a point on the circle of contact;
for the line through V touching any curve traced on the
sphere is a tangent line to the sphere.

Hence the following :—

Construction. Draw a vertical plane through V and
cutting the sphere : let vs be the horizontal trace of that
plane. Now let the plane Vvs be turned about Vv till it
is parallel to the vertical plane of projection, so that the
section of the sphere is projected into the circle $r_1'q_1's_1'$.
From v' draw the tangents $v'p_1'$ and $v'q_1'$ to that circle, touching it at the points p_1' and q_1'. Next let the plane be
returned to its original position, and find the projections
pp' and qq' of the points of contact of the tangents : these
are clearly the projections of two points on the required
circle.

This operation may be repeated with other cutting planes
till a sufficient number of points have been obtained for
tracing the curves.

The projections of the circle are, in general, ellipses,
but if V and O are on the same level the horizontal projection is a straight line, and if they are in the same
vertical line the horizontal projection is a circle : similarly,

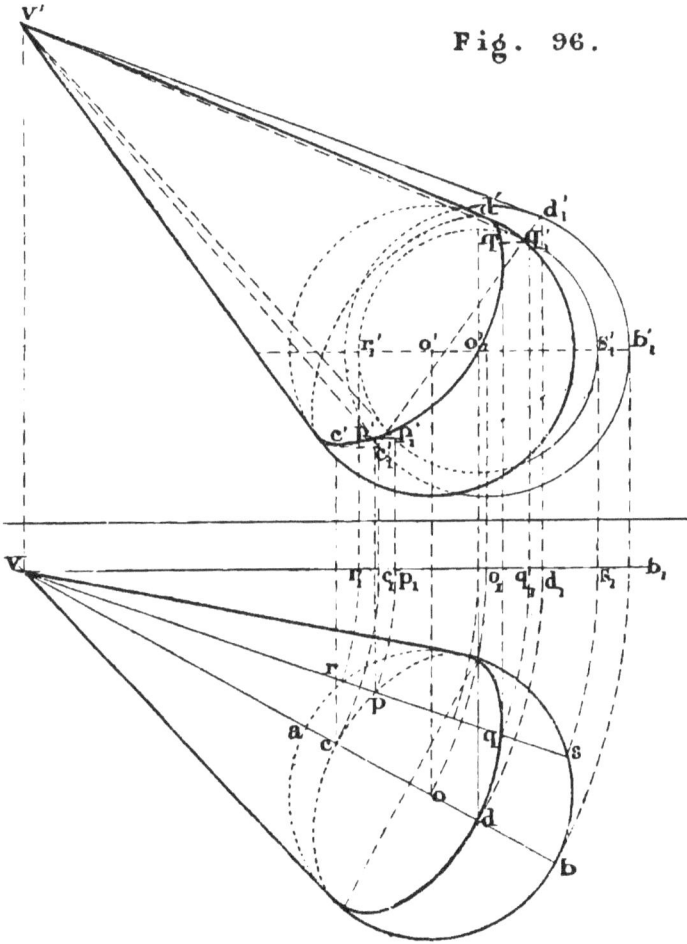

Fig. 96.

the vertical projection may be a straight line or a circle according as VO is parallel or at right angles to the vertical plane of projection.

The axes of these ellipses may be found in the following way, and the ellipses drawn by any of the well-known methods.

As has been shown at page 145 the cone is a right circular one having VO for its axis; and therefore the plane of the circle of contact is at right angles to VO. Hence the vertical plane containing VO cuts the base of the cone, which is the circle of contact, along that diameter which has the greatest inclination, and the projection of which is the minor axis of the ellipse.

Draw, therefore, the vertical plane VvO as in the former case and determine the points c and d. The line cd is the minor axis of the ellipse and the major axis is equal to $c_i'd_i'$ which is the diameter of the circle of contact.

By taking a plane containing VO and at right angles to the vertical plane, the axes of the ellipse which is the vertical projection of the circle may be determined in a similar way.

PROBLEM XVIII.   Fig. 97.

To draw a plane containing a given line and touching a given sphere.

Let AB be the given line, and C the centre of the given sphere, the radius of which is equal to ch; it is required to draw a tangent plane to the sphere which shall contain AB.

Construction.   Draw the traces of a plane LMN containing C, and perpendicular to AB; and find B, the point of intersection of AB and LMN...(Probs. XVI. and XIV. Ch. II.)

Let the plane LMN be rabatted on the horizontal plane of projection, and determine the rabatment of the great circle in which LMN cuts the sphere; that is to say, find C_1 and draw a circle with that centre and a radius equal to ch; also find B_1, the rabatment of the point B, and from B_1 draw $B_1 D_1$ touching the circle at D_1.   Now let LMN be restored to its original position, and find dd', the projections of the point D of which D_1 is the rabatment. D is the point of contact of a plane containing AB and touching the sphere. RST are the traces of that plane, which contains the two lines AB and BD.

Proof.   Because AB is perpendicular to LMN it is perpendicular to CD which lies in that plane.   Also the tangent BD is perpendicular to CD.   Therefore the plane RST which contains AB and BD is perpendicular to the radius of the sphere CD, and as it passes through the point D on the surface it is a tangent plane to the sphere.   But it also contains AB, and is therefore the tangent plane required.

Remarks.   As two tangents $B_1 D_1$ and $B_1 E_1$ can be drawn to the rabatted circle there are evidently two planes which contain AB and touch the sphere. E is the point of contact of the second plane, which contains the two lines AB and BE.

This problem is sometimes solved by drawing two cones having their vertices in the given line, and each enveloping the given sphere.   Each of the cones touches the sphere along a circle, and the two points of intersection of these circles are the points of contact of the two planes which contain the line and touch the sphere.   Each plane is a common tangent plane to the two cones.

When this method is adopted it is convenient, if possible, to take the vertex of one of the cones so that the circle of contact may be projected into a straight line, that is either on the same level with the centre of the sphere or in the

Fig. 97.

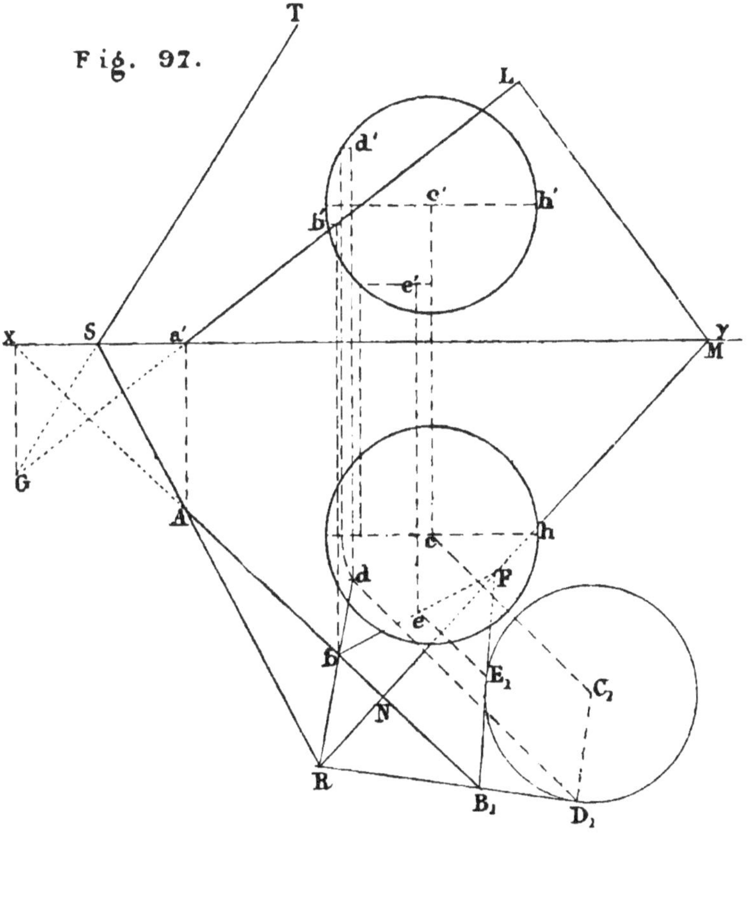

PROBLEM XIX.   Fig. 98.

To find the vertex of a cone which shall envelop two given spheres.

Let *C* and *O* be the centres of the given spheres, and *agb*, *dhe* their horizontal projections.

Construction.   Draw *be*, the common tangent to the two circles *agb*, *dhe*, and find *V* the point of intersection of the line joining the centres of the two spheres, that is *CO*, with the vertical plane having *be* for its horizontal trace......
(Prob. XIV. Chap. II.).   *V* is the point required.

Proof.   The vertical plane containing *be* evidently touches the projecting cylinders of the two spheres along the lines *Bb*, *Ee*, and must therefore touch the two spheres at the points *B* and *E*, respectively.

As the radii *BC* and *EO* are both at right angles to the tangent plane *bvV* they lie in the same plane (Theor. VII. Chap. I.), which is the plane containing the line of centres *CV*.

But the plane *CBV* cuts each of the spheres in a great circle, and cuts the plane *bvV* in the straight line *BV* which passes through *E* and is the tangent to each of these great circles.

Now, if the plane *CBV* were to revolve about the line *CV* the great circles would generate the spheres, and the line *VB* a cone touching them.

Therefore a cone having the vertex *V* and enveloping one of the given spheres, must also envelop the other.

Remark.   There is obviously a second solution to this problem; for the common tangent to the two circles *agb* and *dhe* might have been drawn to intersect the line of centres at the point *w* between the circles, and it may be proved, as before, that the point *W* is the vertex of a conical surface such that if it envelops one of the spheres, it must also envelop the other.

Fig. 98.

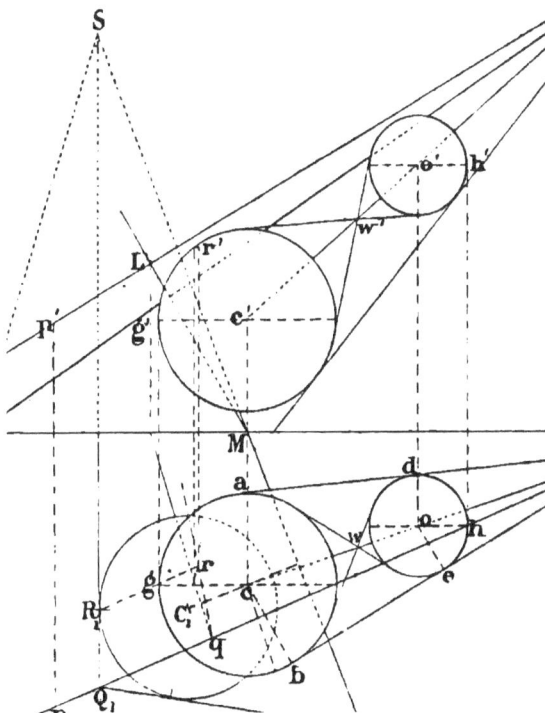

PROBLEM XX.   Fig. 98.

To draw a plane to touch two given spheres and contain a given point.

Let C be the centre of a sphere of radius ca, and O the centre of a sphere of radius od; it is required to draw a common tangent plane to these two spheres which shall also contain the point P.

Construction. Find V, the vertex of the cone which envelops the two given spheres (Prob. XIX.), and draw the traces of a plane TUF containing the line PV and touching one of the spheres (Prob. XVIII.); TUF is the plane required.

In the figure, LMN is a plane containing C and perpendicular to PV; Q is the point of intersection of PV and LMN; Q_1 is the rabatment of Q; and R is the point in which TUF touches the sphere.

Proof. As the plane TUF contains the point V, and touches the sphere (C) in the point R, it contains one of the generators VR of the conical surface which envelops the two spheres; therefore TUF touches the other sphere also; and as it contains P it is the plane required.

Remarks. As in Prob. XVIII. there are two planes which contain the line PV and touch one of the spheres (TN is the horizontal trace of the second plane); and as two conical surfaces may be drawn to envelop the two spheres, there are in all four planes which fulfil the conditions of this problem. The others may be found in a similar way.

It may be observed that the problem might also be solved by drawing a plane containing the point P and touching the conical surface which envelops the spheres, as in Prob. III.

PROBLEM XXI.

To draw a tangent plane to three given spheres.

Let the three spheres be denoted by *A*, *B* and *C*. Then if the vertex of a cone enveloping *A* and *B* be determined, and also the vertex of a cone enveloping *A* and *C*, or *B* and *C*, it follows from Prob. XX. that the plane which contains these two vertices and touches one of the spheres must also touch the other two.

Now there are two conical surfaces which envelop *A* and *B*, and two which envelop *A* and *C*, and two which envelop *B* and *C*; that is six altogether. The six vertices of these cones will be found to lie on four straight lines, three on each line; and two planes can be drawn containing any one of these lines, and touching the spheres; so that in all eight planes may be drawn to touch three given spheres.

HYPERBOLOID OF REVOLUTION.

PROBLEM XXII. Fig. 99.

To find the projections of the surface generated by a straight line which revolves about an axis not in the same plane with it.

Let the horizontal plane of projection be taken at right angles to the axis, the projections of which are o, and $n'a'$; and let the initial position of the generating line be parallel to the vertical plane of projection, so that aB, $a'b'$ are its projections.

The generator may be considered as indefinitely extended, but in the figure the solid is bounded by the horizontal plane of projection and the plane described by the common perpendicular of the axis and generator—that common perpendicular is the line OA, equal and parallel to oa.

As the axis is perpendicular to the horizontal plane the horizontal trace of the surface will be the circle described by the horizontal trace of AB; that is the circle through B described from the centre o.

The outline of the vertical projection will be the projection of the meridian parallel to the vertical plane; that is the curve of intersection of the generator with the meridian plane F_1B_1. It is required to find that curve.

Construction. Take any point D on the generator. The distance of D from the axis is equal to od; and as the generator revolves about Oo, the point D moves on the circumference of a circle. Find the vertical projections of the points D_1 and H_1 in which the path of D intersects the meridian plane F_1B_1; h_1' and d_1' are points on the required curves. In the same way any number of points, as c_1', e_1' and g_1', k_1', may be found, and the curve traced through them. A is the point of the generator nearest to the axis, and its path AA_1A_2 is consequently the smallest parallel of the solid. It is called the throat circle.

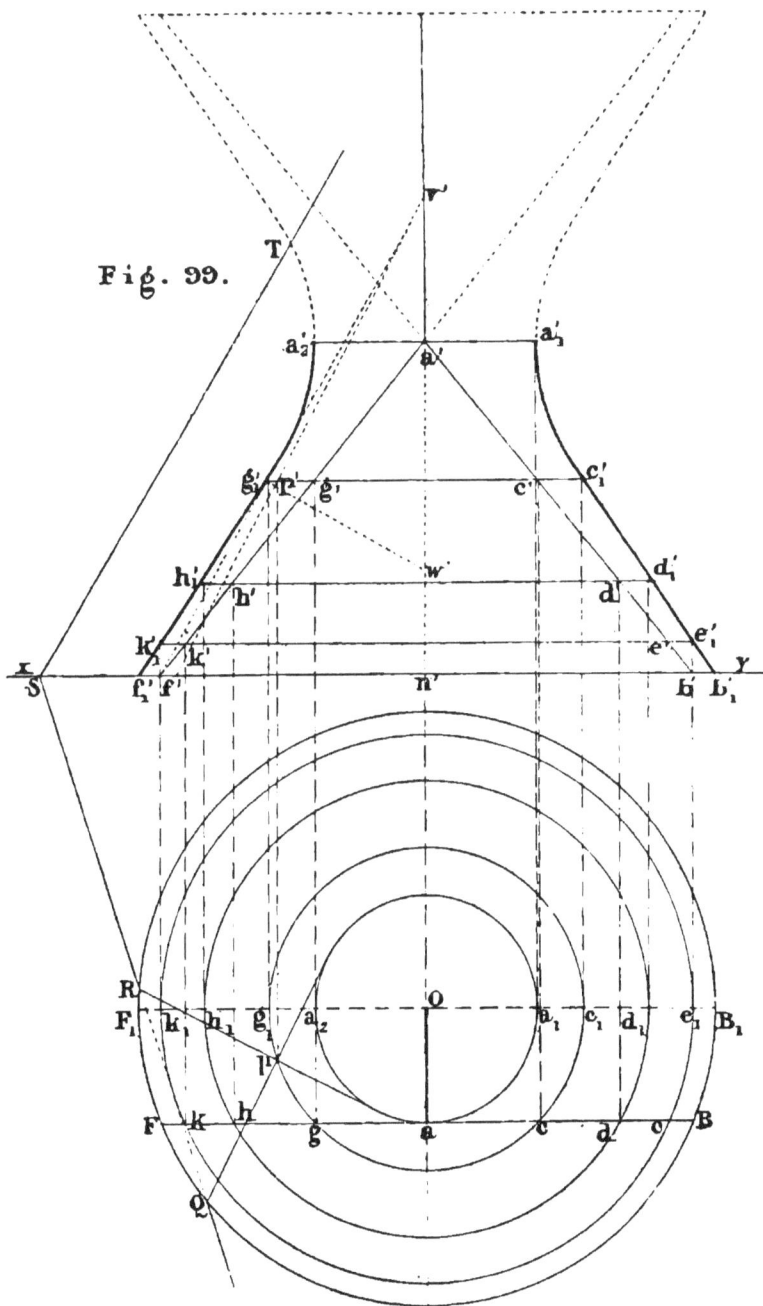

Fig. 99.

Notes. 1. The meridians of the surface are hyperbolas, and as the surface might be generated by one of these hyperbolas revolving about the axis, it is called the *Hyperboloid of revolution.*

2. As the line *OA* is always horizontal and at right angles to the generator, the horizontal projections of these two lines are at right angles to one another in all their positions. (Theor. XVII. Ch. I.) Therefore *the horizontal projection of the generator is always a tangent to the horizontal projection of the throat circle.*

3. It is evident that the same surface might be generated by the straight line *AF*, which has the same inclination to the axis as *AB* but in the opposite direction. Hence, *through any point of the surface, two straight lines can be drawn which coincide with the surface throughout their whole length.*

PROBLEM XXIII.   Fig. 99.

To draw a tangent plane to a hyperboloid of revolution at a given point.

Let p be the horizontal projection of a point on the surface; it is required to find the traces of a tangent plane to the surface at P.

Construction. Find p' as in Problem XIII. Through p draw pQ and pR tangents to the circle a_2aa_1. pQ and pR are the horizontal projections of the two generators of the surface which pass through the point P. Determine the traces QST of the plane containing the two lines PR and PQ. QST is the plane required.

Proof. QST contains the two straight lines passing through P and coinciding with the surface throughout their whole length; it is therefore a tangent plane at the point P. (Theorem I.)

Notes. 1.   The tangent plane QST should contain the tangent to the parallel through P.

2.   The tangent plane at P cuts the axis at V, so that V is the vertex of a cone which touches the hyperboloid along the parallel G_1PC. $v'g_1'$ is the vertical trace of the tangent plane at the point G_1. W is the vertex of the normal cone along G_1PC.

3.   It is evident that at different points of the line PQ, there are different tangent planes, so that, unlike the cone and cylinder, the tangent plane to a hyperboloid of revolution does not touch the surface along a generator. In fact it touches at one point only, and cuts the surface along the two generators passing through that point.

PROBLEM XXIV.   Fig. 100.

To find the plane section of a hyperboloid of revolution.

Let the vertical line through o be the axis; AB or AF the initial position of the generator; and LMN the traces of a plane; it is required to draw the projections of the curve of intersection of LMN with the hyperboloid generated by AB or AF revolving about the axis Oo.

Construction. Draw a horizontal plane QR intersecting the plane LMN in a line, of which the projections are QR and ru, and intersecting the generator AF in a point S. As the generator revolves, the point S moves on the circumference of a circle, the horizontal projection of which is sut, and the vertical projection coincides with QR. Now as the circle SUT lies on the surface of the hyperboloid and the line RU lies in the given plane, the points U and T in which the line and circle intersect must be two points in the curve required; uu' and tt' are therefore the projections of two points in the curve. In a similar way the projections of any number of points in the curve may be determined, and the projections of the curve drawn through them. There are, however, some points which it is best to determine by a special method, such as G and H, the vertices of the curve (it is evident the curve must be symmetrical with respect to LN, which is consequently an axis), or the points α', β', where the vertical projection of the curve touches the hyperbolas.

To find the projections of the points G and H. Find d', the vertical projection of the point in which LMN cuts the axis of revolution. The points G and H are evidently the points of intersection of the generator and the line DL. Now a little consideration will show that in order to find the circle described by the point of intersection it will be the same whether the line DL is fixed and AF revolving, or AF fixed and DL revolving about the same axis Oo. Conceive the line DL to revolve; it will generate a cone of which the

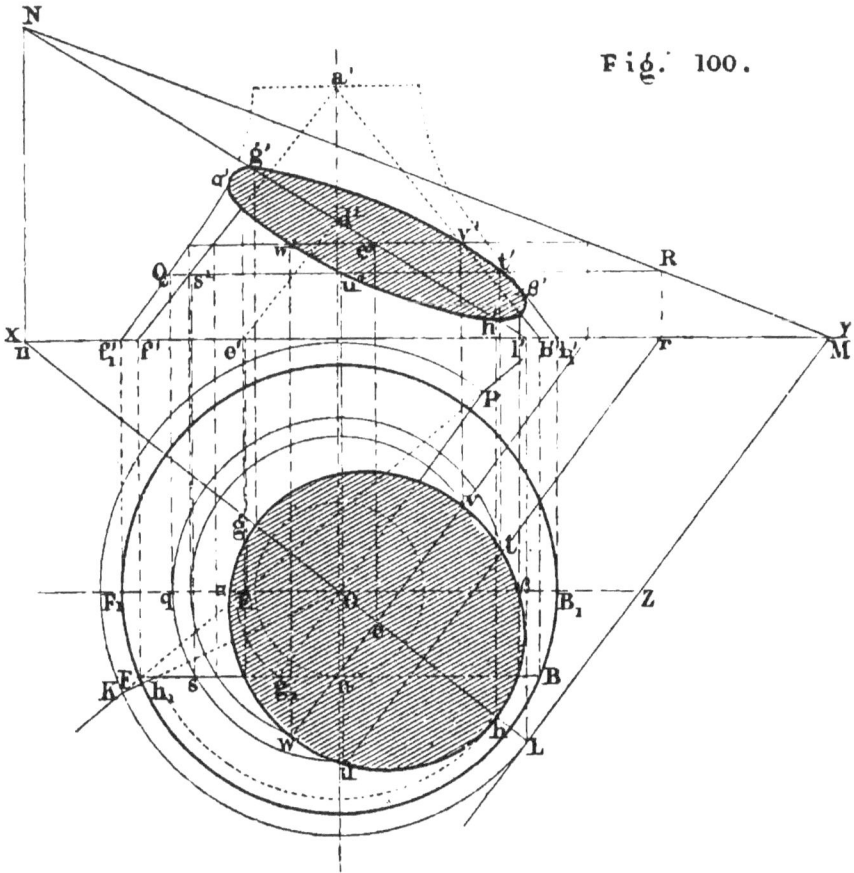

Fig. 100.

horizontal trace is LKP. To find the point in which AF intersects this cone, draw the horizontal trace of the plane containing AF and the point D; KP is that trace. As the generator DK of the cone and the line AF are in the same plane, h_1 is the horizontal projection of their point of intersection. Let DK revolve about Oo till it coincides with DL; h_1 comes to the position h, which is one of the points required. A second generator PD of the cone cuts AF in a point of which g_1 is the horizontal projection, so that making og equal to og_1 the horizontal projection of G is found.

Similarly, by conceiving the line DZ in the plane LMN to describe a cone about Oo, and finding the points in which AF meets two generators of that cone, the points α' and β' may be found. α' and β' lie, of course, on the vertical projection of DZ.

Note. If from O, the point in which the common perpendicular of the axis and generator meets the axis, a line be drawn parallel to AB, and if that line revolves with AB, always remaining parallel to it, it generates a cone which is called the asymptotic cone of the hyperboloid. $a'f'b'$ is its vertical projection, and it is such that in any meridian section of the two surfaces, the straight lines which are the sections of the cone are asymptotes to the hyperbolas which are the sections of the hyperboloids.

The following theorem which can be proved by analytical geometry may here be stated.

Theorem. The sections of a hyperboloid of revolution and its asymptotic cone, by the same plane, are of the same kind and have a common centre.

It is to be borne in mind that a point, straight line, or pair of intersecting straight lines, are the limits of the ellipse, parabola, and hyperbola, respectively.

1. A right circular cone, height 3″ and diameter of base 2″, rests with its base on a plane of which the vertical and horizontal traces make angles of 30° and 45° respectively with the ground line. Draw its plan, elevation, and traces.

2. Draw a plane inclined at 60° and touching the cone of Ex. 1.

3. Draw the plan of a cube of 2″ edge when two adjacent faces are inclined at angles of 50° and 70° respectively.

4. The height of a regular hexagonal pyramid is 3″, and the side of its base 1″. Draw its plan when the base is inclined at 60° and one of the triangular faces inclined at 70°.

5. The base of a cone of revolution is 3″ diameter, height 4″. Represent the true form of the section by a plane which cuts a generator one inch from the vertex—that being the highest point of the section:

(1) When the plane cuts the axis at an angle of 60°.

(2) When the plane is parallel to one of the generators.

(3) When the plane is parallel to the axis.

6. Draw the developments of the sections of the last exercise.

7. Take any two points on a cone, not on the same generator, and show the plan and elevation of the shortest line which can be drawn between them on the surface of the cone.

8.   The axis of a right circular cylinder of 2″ diameter is inclined at 45° to the horizontal and parallel to the vertical plane of projection.   Determine the two planes touching the cylinder and inclined at 60° to the horizontal.

9.   Find the horizontal trace of this cylinder (Ex. 8) and draw its development.

10.   On a rectangular slip of paper, 6″ long, a line is drawn making an angle of 30° with the edge; the paper is then wrapped on a cylinder of 6″ circumference.   Draw the projection of the line on a plane parallel to the axis of the cylinder.

11.   A surface is generated by a straight line moving in contact with the helix and axis of the cylinder of the last example, and at right angles to the axis.   Determine the section of that surface by a plane making an angle of 60 with the axis.

12.   A sphere of 2″ diameter rests on the horizontal plane, and is touched by three planes which are equally inclined to one another and at 50° to the horizontal.   Find the points of contact, the height of the pyramid so formed, and the inclinations of the tangent planes to one another.

13.   Three spheres of 1″, 1½″ and 2″ diameter respectively lie on the horizontal plane, each sphere touching the other two.   Draw their plan and elevation, and the traces of a plane, not horizontal, touching them all.   Mark the points of contact.

14.   A sphere of 2″ diameter has its centre raised 3″ above the horizontal plane; the axis of a cylinder enveloping the sphere is inclined at 45°.   Find the line of contact of the two surfaces and the horizontal trace of the cylinder.

15.   Determine the surfaces generated by the edges (2″ long) of a cube when it revolves about one of its diagonals.

CHAPTER V.

THE general method of determining the line of intersection of two surfaces is to take a series of auxiliary surfaces, which are mostly plane, cutting both surfaces, these auxiliary surfaces being so chosen as to cut the given surfaces in lines which are easily drawn, such as straight lines or circles. The point in which a line on one surface intersects a line on the other is common to the two surfaces and consequently a point in their curve of intersection. For instance, let the two given surfaces be denoted by S and S'', and let a plane P cut S in the line L and S'' in the line L'; then if the two lines L and L' intersect in the point M, that is a point in the required curve. The projections of a sufficient number of such points being found, the projections of the curve of intersection of the two given surfaces may be traced through them. Examples of this method have been given in Chap. IV. in finding the plane sections of curved surfaces, in other words in finding the common section of two surfaces, one being curved and the other plane. Thus in finding the plane section of a hyperboloid of revolution a series of planes were drawn perpendicular to the axis, so as to cut the surface in circles, and the points of intersections of any one of these circles with the straight line in which the auxiliary cut the secant plane gave two points in the curve required.

The line of intersection of two curved surfaces is in general a *curve of double curvature*, that is a curve all the points of which are not in the same plane.

The tangent to the curve of intersection of two curved surfaces at any point, is the common section of the tangent planes to the surfaces at that point; for the curve being at the same time on both surfaces its tangent must lie in both the tangent planes.

PROBLEM I. Fig. 101.

To find the curve of intersection of two given conical surfaces.

Let *ABC* be the horizontal trace of a cone and *v v'* the projections of its vertex; *DEF*, the vertical trace of a second cone and *w w'* the projections of its vertex; it is required to find the projections of the curve in which the two surfaces intersect.

Construction. If auxiliary planes be taken which pass through the vertices of the cones they will cut them in straight lines. Draw, therefore, the projections of the line *VW*, and find its traces *T* and *U*. Any plane containing *TU* passes through the vertex of each cone. Draw *UH* cutting the horizontal trace of the first cone in the points *B* and *C*; and join *H* with *T*, the line *HT* cutting the vertical trace of the second cone in the points *E* and *F*. Now the plane *THU* cuts the first cone along the two straight lines *BV*, *CV*, and cuts the other cone along the straight lines *EW*, *FW*. The four points *K*, *L*, *M*, *N*, in which these lines intersect, are four points in the curve required. *k'*, *l'*, *m'*, *n'*, are the vertical projections of these four points and *k*, *l* the horizontal projections of *K* and *L*, the other two being omitted in plan for the sake of clearness of the figure. In the same way, by drawing any other plane containing *TU* and cutting the two surfaces, four more points may be found. A few of these planes are shown on the figure.

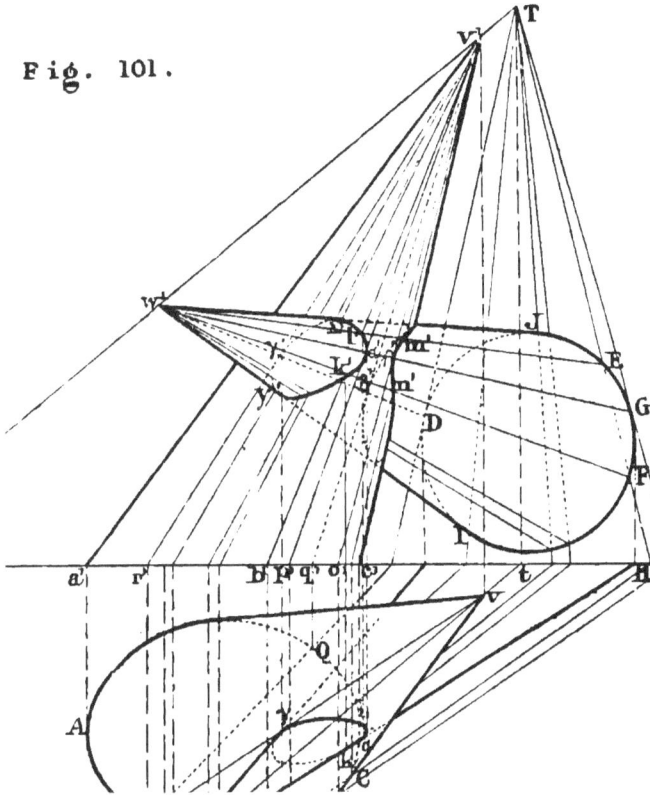

Fig. 101.

One of the auxiliary planes touches the second cone along the line GW and cuts the first cone along the two lines OV, PV. In this case the four points are reduced to two, $\alpha\alpha'$ and $\beta\beta'$, as may be seen from the figure. Again, there is a second tangent plane touching the second cone along DW and cutting the first cone, which also gives two points of the curves. These are called the *limiting planes*, for the curves of intersection must lie wholly between them.

To find the point situated on any given generator of either cone, take an auxiliary plane containing that line and find the point in the same way as before. For instance, to find the point on the generator WJ, that is the highest line of the cone, of which $w'J$ is the vertical projection ; TJ is the vertical trace of the plane containing that line, and determines the point X on the curve. It is important to find in this way the points on the contour or outline of the projections of the cone, for, as may be easily seen, these lines are tangents to the projections of the curve. Thus $w'J$ is a tangent to the curve $k'l'x'$ at the point x'. Similarly the line lw' touches the same curve at y'.

Remarks. Particular attention has been directed to the vertical projection of the curve in this problem, but the horizontal projection may be found in a similar way, by finding the points of intersection of the generators of the surfaces which are in the same plane, or when the projections of the generators intersect very obliquely, as the horizontal projections do in the figure, one projection of the curve may be found from the other by drawing the generators of one of the cones.

When the limiting planes both cut the same surface, as in this example, there are two curves of intersection. The smaller cone penetrates the larger and is completely enclosed by it, so that there is a curve of entrance and one of emergence. If both the limiting planes were not tangents to the same surface the two curves would merge into one another and form one curve.

As the tangent plane to the second cone along the line GW cuts the first cone along the generator PV, that generator is the common section of the tangent planes to the two cones at the point of intersection of GW and PV, and consequently a tangent to the curve of intersection. As the same may be said of the other points of the curve which are situated on the limiting planes, it may be stated generally, that *the tangents to the curve of intersection at a point situated on a limiting plane is the generator in which that plane cuts the other cone.*

PROBLEM II.   Fig. 102.

To find the curve of intersection of a cone and a cylinder.

Let ABP be the horizontal trace of a cone, and V its vertex; CDO, the horizontal trace of a cylinder, and KL one of its generators; it is required to find the curve in which the two surfaces intersect.

The planes which cut the cylinder in straight lines, that is along generators, must be parallel to KL; while planes that cut the cone in straight lines pass through the vertex. Hence the auxiliary planes are to be taken passing through V and parallel to KL. In fact this problem may be considered as a particular case of Prob. I., the vertex of one of the cones being at an infinite distance.

Construction. Through the point V draw VT parallel to KL, and find its horizontal trace T. Now planes containing VT fulfil the required condition of cutting both surfaces in straight lines, so that any line through T, cutting the horizontal traces of the surfaces, may be taken as the horizontal trace of an auxiliary plane.

Draw any straight line AT cutting ABP in the points A, B, and CDO in the points C, D. The plane ATV cuts the cone along the two straight lines AV, BV, and the cylinder along the two straight lines CE, DF. AV meets CE in the point E, and DF in the point F; BV meets CE in the point G, and DF in the point H. Therefore E, F, G, H are four points in the curve required.

To find the points on the limiting planes, draw TK and TO tangents to CDO and cutting ABP at I, J, P, Q. These limiting planes touch the cylinder along the straight lines KL and OS, and cut the cone along the four lines IV, JV, PV and QV. The four limiting points of the curves are L, M, R and S. The lines IV, JV, PV and QV are tangents to the curves at the points L, M, R and S respectively, as has been shown in the preceding problem. To find the points of the curve which are on any given generator of either surface, draw the auxiliary plane containing that generator and proceed as before. For instance, to find the points on $u'\beta'$, draw TU cutting ABP at X and Y; the points of intersection of $x'v'$ and $y'v'$ with $u'\beta'$ are the points where the curves meet $u'\beta'$.

Fig. 102.

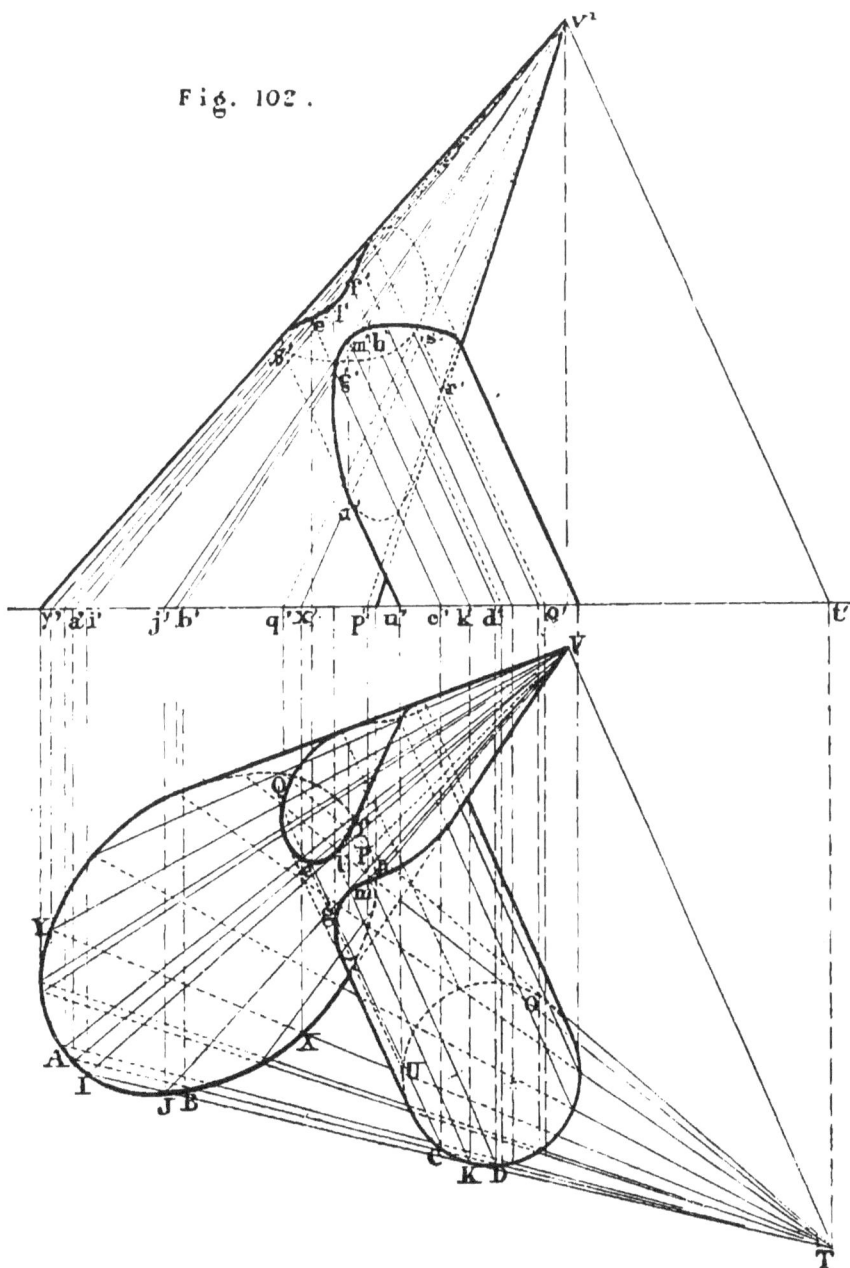

PROBLEM III. Fig. 103.

To find the curve of intersection of two cylinders.

Let *ABO* be the horizontal trace of a cylinder, and *Pv*, *p'v'* the projections of one of its generators; *CDN* the horizontal trace of a second cylinder, and *Km*, *k'm'* the projections of one of its generators; it is required to find the projections of the curve in which the two surfaces intersect.

As the section of a cylinder by a plane parallel to its generators is one or more straight lines, the auxiliary planes must be taken parallel to the generators of both the given surfaces.

Construction. From any point *V* of the generator *PV* draw *VT* parallel to *KM*, and find *T* its horizontal trace. Now the plane *PVT* is parallel to the generators of both cylinders (Theor. X. Chap. I.). Consequently the auxiliary planes are to have their horizontal traces parallel to *PT*.

Draw any straight line *AD* parallel to *PT*, cutting the horizontal trace of one cylinder in the points *A*, *B*, and the other in the points *C*, *D*. The points *A*, *B*, *C* and *D* are the horizontal traces of four generators, two on each surface, which lie in the same plane, so that their points of intersection *E*, *F*, *G* and *H* are four points of the required curve. In a similar way as many points as required may be found on the curve.

The limiting planes are *JIL* and *NQS*. As these planes are not both tangents to the same surface, neither cylinder is enclosed entirely within the other. In the figure one of the cylinders is supposed to be removed.

Fig. 103.

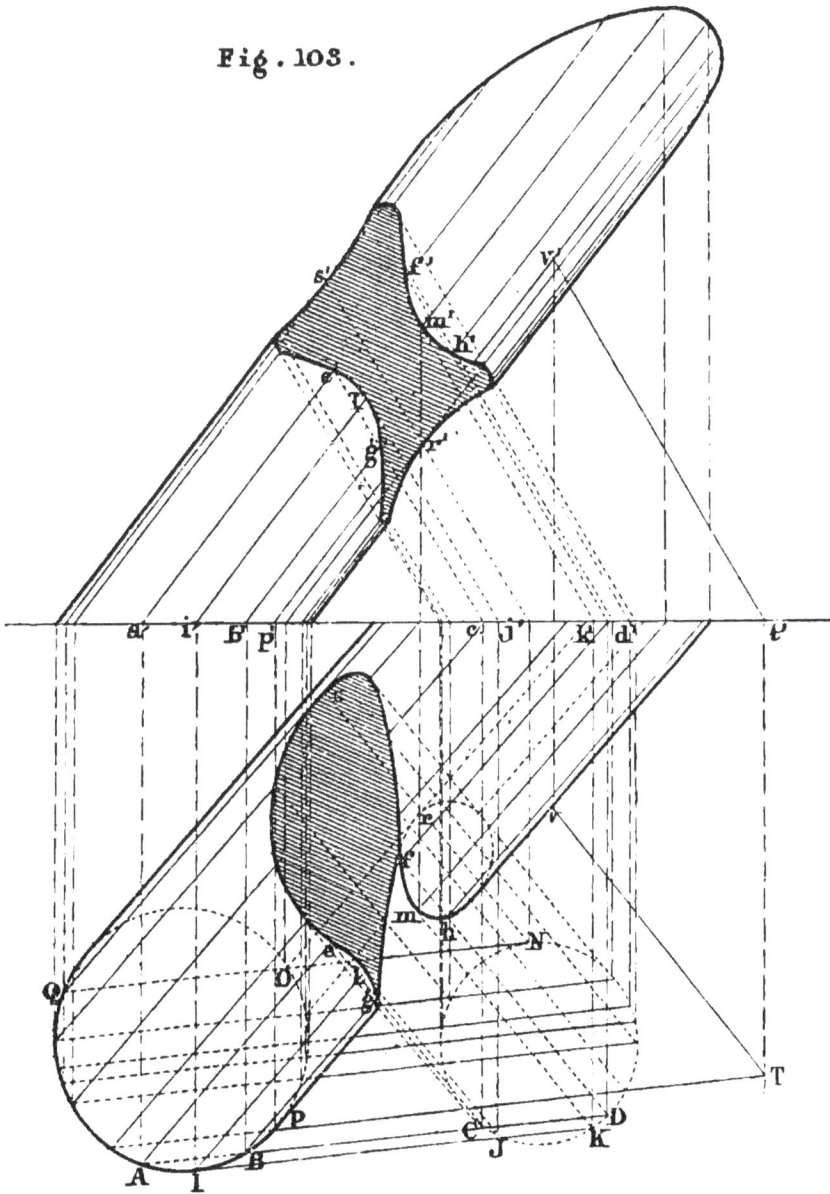

PROBLEM IV. Fig. 104.

To find the curve of intersection of a cone and a right circular cylinder when the axis of the cylinder is parallel to the ground line.

Let $PQRS$ be the horizontal trace of the conical surface and V its vertex; CD the axis of the cylinder, and its radius equal to ce.

Any plane parallel to the generators of the cylinder, and consequently to its axis, must have its horizontal trace parallel to the ground line (Theor. XI. Cor. 2. Ch. I.). So that any line AB parallel to the ground line and intersecting the horizontal trace of the cone may be taken for the horizontal trace of an auxiliary plane, provided it also meets the cylinder.

To find if the plane VAB cuts the cylinder, and, if it does, to determine the lines of section. Conceive a plane perpendicular to CD to pass through V, cutting the cylinder in a circle and the plane VAB in a straight line VH_1. If this straight line and circle intersect one another their points of intersection are on the cylinder, and consequently points on the generators along which VAB cuts the surface. Now, if the plane VH_1v were turned about Vv as an axis till it came parallel to the vertical plane of projection, H_1 would come to H, and its vertical projection would be $h'v'$, and the circular section of the cylinder would have for its vertical projection a circle described about o', on the axis of the cylinder, with a radius equal to the radius of the cylinder. As the line $v'h'$ cuts the circle in the two points f'', g', the plane VAB cuts the cylinder along the two lines of which $f''m'$ and $g'n'$ are the vertical projections. The points on the curve which are in the plane VAB are the points of intersection of FM, GN, AV and BV. In working the problem, however, it will be better to proceed as follows.

Construction. Draw vH_1 for the horizontal trace of a vertical plane, which cuts the axis of the cylinder at right

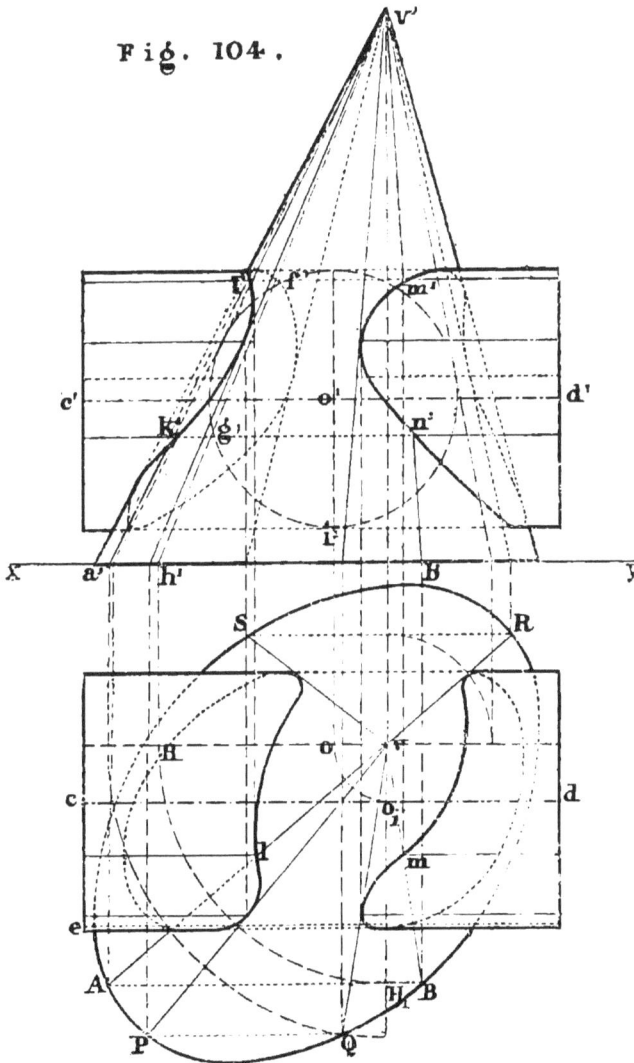

Fig. 104.

angles in the point O_1. Let this plane be turned about Vv
as an axis till it is parallel to the vertical plane of projection
and the point O_1 comes to O. From o' as centre and $o'i' = ce$,
draw the circle $f'g'i'$. Now from v' draw any line $v'h'$ cutting
the circle in the points f' and g'. With centre v and radius
vH describe a circle cutting vH_1 at H_1, and draw AB through
H_1 parallel to xy. AB is the horizontal trace of an auxiliary
plane which cuts the cone along the two lines AV, BV, and
the cylinder along the two lines FM, GN. F, G, M and N
are consequently four points on the required curve.

The limiting planes are, as in former cases, those which
touch one surface and cut the other. In this example they
both touch the cylinder, and are determined by lines drawn
from v' touching the circle, as may be seen from the figure.
They are the two planes PQV and RSV.

12—2

PROBLEM V. Fig. 105.

To find the curve of intersection of a cylinder and a surface of revolution.

GRF is the horizontal trace of the cylinder, and *FII* one of its generators; the vertical line through *O* is the axis of the surface of revolution, and $o'a'p'q'$ the projection of that meridian which is parallel to the vertical plane of projection.

This problem might be solved by taking horizontal auxiliary planes, cutting the surface of revolution in circles and the cylinder in curves equal and similar to its horizontal trace. But the construction of these latter curves is, in general, too laborious, and the following method will be found preferable.

Construction. Take any plane section *ABC* of the surface of revolution, at right angles to its axis. Let this circle be taken as a directrix of a cylinder, having its generators parallel to those of the given cylinder. The horizontal trace of that auxiliary cylinder is the circle *GDF*, equal to *ABC*. As the horizontal traces of the given and auxiliary cylinders intersect at the points *G* and *F*, the two cylinders intersect along the two straight lines *FII* and *GK*, and the points *H* and *K* where these two lines meet *ABC* are evidently points on both the given surfaces, and consequently points on the required curve.

In a similar way as many points as desired may be found on the curve. It is clear there are two auxiliary cylinders which touch the given one, namely, those of which the horizontal traces touch *GRF*. These give the limiting points of the curve, but in general they can only be determined by trial.

Note. In finding the intersection of a cone and a surface of revolution, the auxiliary surfaces are to be cones having the same vertex as the given one, and circular sections of the surface of revolution for their directrices. The horizontal traces of these auxiliary cones will be circles, and the construction will be similar to that given above.

Fig. 105.

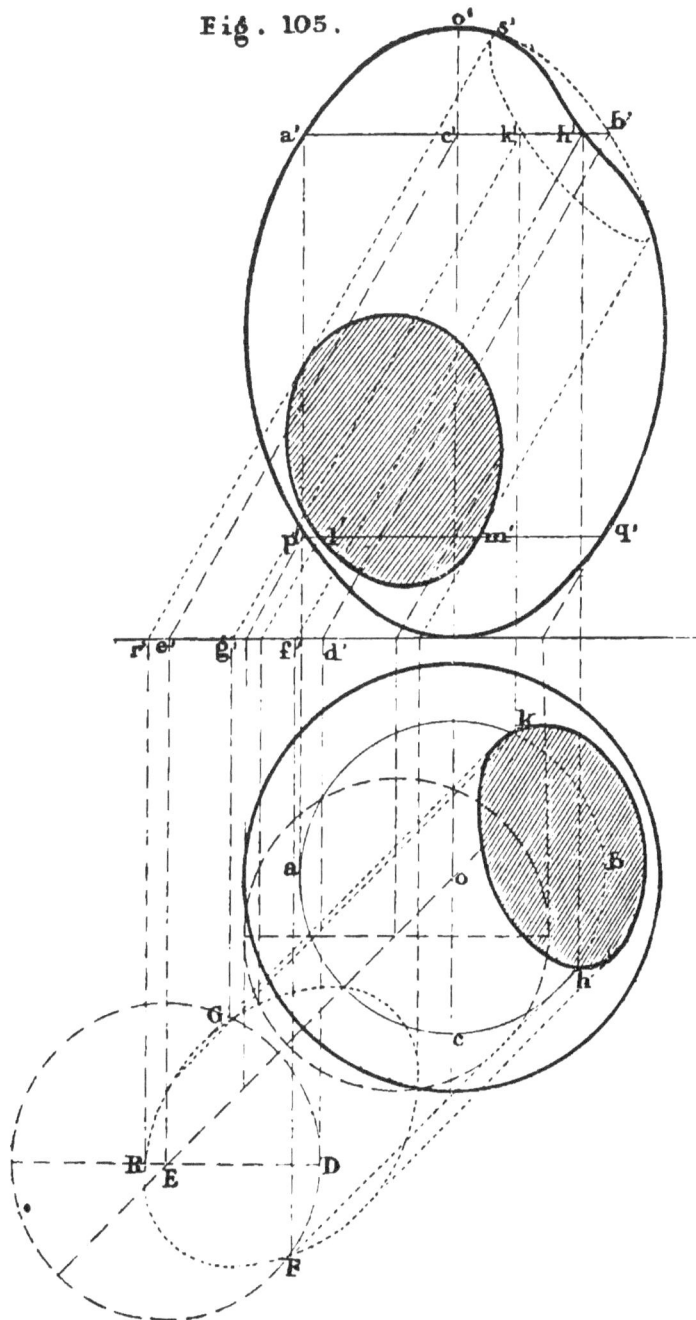

PROBLEM VI. Fig. 106.

To find the curve of intersection of two surfaces of revolution when the axes are parallel.

As the axes are parallel to one another, they are both at right angles to the same plane (Theorem VIII. Ch. I.), so that the auxiliary planes may be taken perpendicular to the axes, and consequently cutting both surfaces in circles.

Let one of the given surfaces be that generated by the circle *IBP* revolving about the axis *Oo*, that is to say, a ring of circular section; and let the other surface be the cone generated by the straight line *VU* revolving about the axis *Vv*. The horizontal plane is, as usual, taken at right angles to the axes.

Construction. Draw any line $a'b'$ parallel to *xy*, and cutting the vertical projections of the two surfaces. This is the vertical trace of a horizontal plane which cuts the cone in the circle *EFG* and the ring in the two circles *AEB, CGD*. The circle *EFG* cuts *AEB* at the points *E* and *F*, and the circle *CGD* at *G* and *H*. Therefore the four points *E, F, G, H* are on the required curve. In a similar way as many points as are necessary for determining the curve may be found.

The limiting points of the curve are those points at which a circular section of one of the surfaces *touches* the circular section of the other. As the points of contact of the horizontal projections of these circles must always be on the line of centres *ov*, it follows that the limiting points of the curves are on the lines *VL* and *VM*, that is, the two generators of the cone which are in the plane of the axes. Hence the limiting planes may be found as follows:—

Let the plane of the axes, together with its sections of the two given surfaces, be turned about *Oo* till it is parallel to the vertical plane of projection. The elevation of the section of the ring is the circle $i'b'q'$, and the elevation of the section of the cone is the isosceles triangle $l'_{,}v'_{,}m'_{,}$. The limiting planes are those of which the vertical traces pass through the points p', q', i', j', and the limiting points are determined from them in the usual way. They are P_{1}, I_{1}, Q_{1}, and J_{1}.

In this figure the cone penetrates the ring, and there are two curves of intersection; the lower one has been omitted in the plan for the sake of clearness. The cone is supposed

Fig. 106.

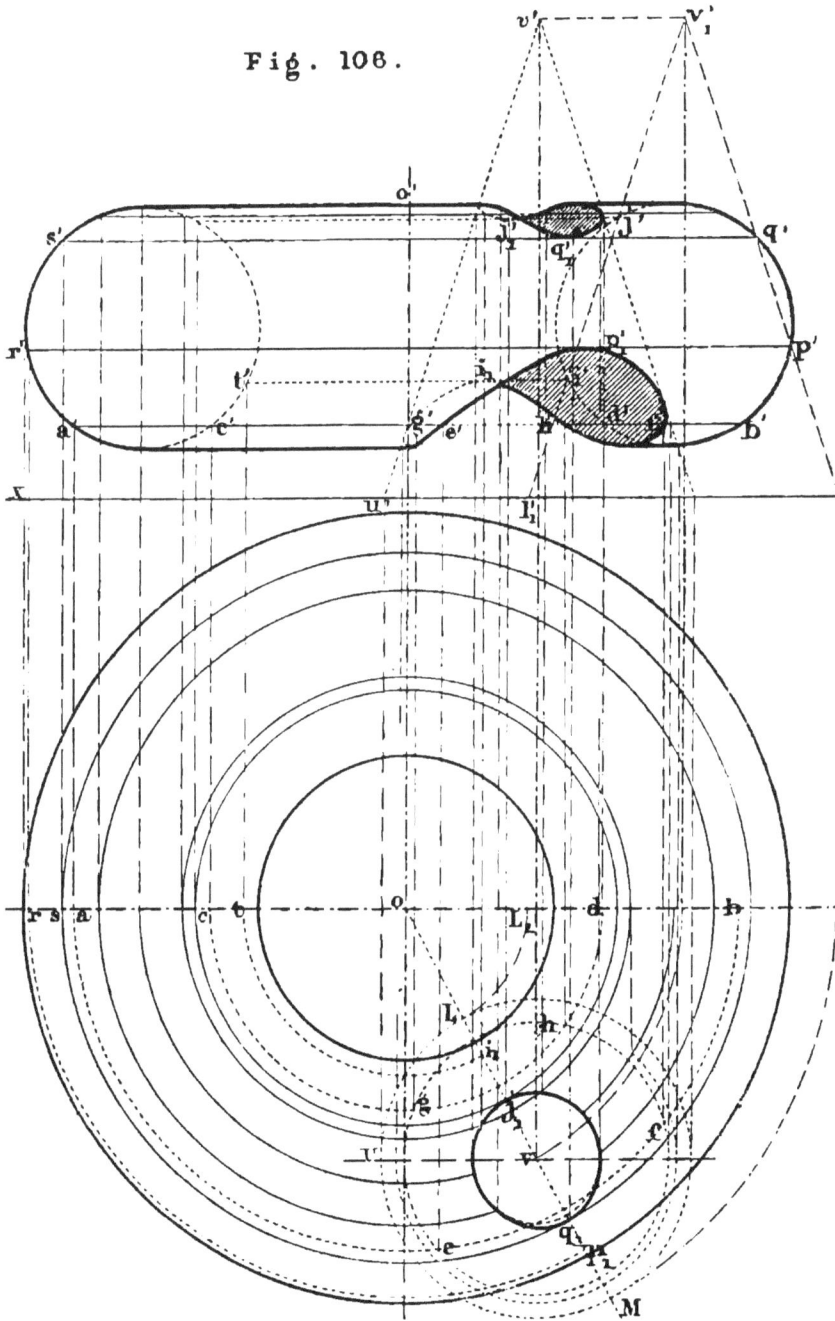

PROBLEM VII.   Fig. 107.

To find the curve of intersection of two surfaces of revolution when their axes intersect.

Let Oo be the axis and PEC the meridian of one surface; OD the axis and PGQ the meridian of the second surface; it is required to find their common section. The vertical plane of projection is taken parallel to the axes, and the horizontal plane perpendicular to one of them.

When two surfaces of revolution which cut each other have the same axis they must intersect in one or more circles; for as every point on either meridian describes a circle, the point of intersection of the two meridians describes a circle about the common axis, which is the intersection of the two surfaces.   Hence, as any diameter of a sphere may be considered as its axis, *when the centre of a sphere is on the axis of a surface of revolution, if the sphere and that surface intersect their common section is a circle which is a parallel of the given surface.*   By taking for auxiliary surfaces spheres having their common centre at the point of intersection of the axes, the given surfaces will be cut in circles by the spheres.

Construction.   From the centre o' describe a circle $e'f'h'$ intersecting the elevations of the two given surfaces.   Suppose $e'f'h'$ to be the elevation of a sphere, it intersects the first surface in the parallel EKF, and the second surface in the parallel GKH.   These two circles are in planes perpendicular to the vertical plane of projection, since the axes are parallel to that plane, and their common section is consequently the straight line through K perpendicular to the vertical plane; that is, the line KL. K and L are points on the curve required.   In a similar way any number of points on the required curve may be determined.

In the figure the surface shown in dotted lines is supposed to be removed.

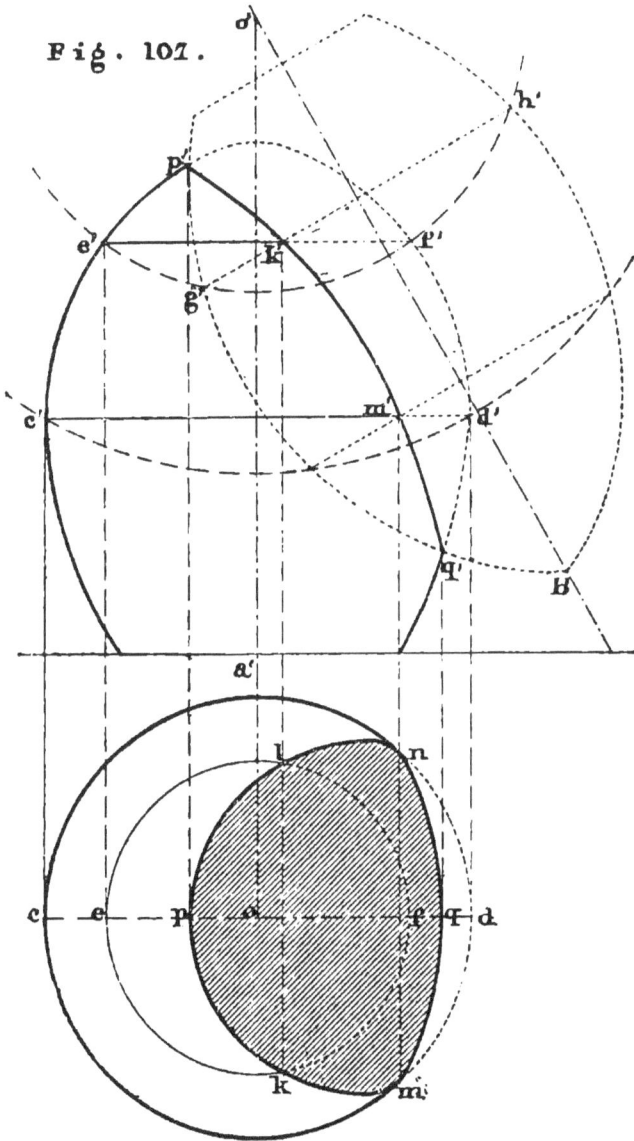

Fig. 101.

It is well known that in a homogeneous medium light is propagated in straight lines. Hence when light from any luminous point falls on an opaque body there is a certain portion of space behind the body as well as part of the body itself which is deprived of the direct rays from the point. The line on the surface of the body which separates the illumined from the dark side is called the line of shade, and the dark space behind is called the shadow of the body. When any surface comes within the shadow so as to have the whole or part of it deprived of the direct rays from the luminous point, it is said to have a shadow *cast* upon it by the opaque body.

By assuming that the medium surrounding the bodies is homogeneous and the source of light a fixed point, the line of shade on any known surface and the shadow cast by it on any other known surface can be determined by Geometry, provided the relative positions of the two surfaces to one another and to the luminous point are given. For the shadow-surface, that is the surface separating the shadow from surrounding space, is generated by straight lines passing through the luminous point and touching the surface which casts the shadow. The points of contact determine the shade line, and the intersection of the shadow-surface with the other given surface is the outline of the cast-shadow. The outline of the cast-shadow is therefore the common section of two known surfaces. For example, in fig. 96, if V were the source of light, the line $CPQD$ would be the line of shade on the given sphere, and the line of intersection of the right circular cone $VCPQD$ with any other surface would be the shadow cast by the sphere on that surface. For instance, the horizontal trace of that cone would be the shadow cast by the sphere on the horizontal plane of projection. It would evidently be an ellipse.

In putting the shadows on Engineering and Architectural Drawings it is the custom to take the rays of light parallel

to one another, which is equivalent to assuming the source
of light at an infinite distance, so that the shadow-surface is
a cylinder, according to the definition in Chapter IV. More-
over the rays of light are always assumed to proceed forward
and downward from the left-hand side—frequently in the
direction of the diagonal of a cube having two of its faces
coinciding with the planes of projection, so that the
projections of the rays make angles of 45° with the ground
line. This direction is found very convenient.

1. A right circular cylinder, 2″ diameter, penetrates a regular hexagonal prism, the shortest diameter of which is 3″: the axis of the cylinder meets the axis of the prism at right angles. Draw the projection of the line of intersection on a plane parallel to the axis of the cylinder and one of the faces of the prism.

2. If the prism of the last example penetrates a sphere of 4″ diameter, the axis of the prism passing through the centre of the sphere, draw the projection of the line of intersection on a plane parallel to a face of the prism.

3. A right circular cylinder of 2″ diameter penetrates another of 3″ diameter, the axes passing $\frac{1}{4}$″ from one another, and at an angle of 60°. Draw the projection of the two solids on a plane parallel to their axes.

4. A cone of revolution, 3″ diameter at base, and 4″ high, has a circular cylindrical hole, $1\frac{1}{2}$″ diameter, bored through it; the axes are at right angles to one another and $\frac{1}{8}$″ apart; the axis of the hole 1″ from the base of the cone. Draw the plan and elevation of the cone when standing on the horizontal plane, the axis of the hole making an angle of 30° with the vertical plane of projection.

5. Draw the development of the conical surface of the last example, showing the holes.

6. Draw the projections of the curve of intersection of a cone and a sphere

(1) When the cone is one of revolution and its axis passes through the centre of the sphere.

(2) When the cone is one of revolution but the axis does not pass through the centre of the sphere.

(3) When the cone is not one of revolution.

7. The longest diameter of the base of a regular octagonal prism is 2″ and its height 4″; it is placed with its base on the horizontal plane, and its axis 2″ from the vertical plane of projection. Find the shadow cast by it on the two planes of projection

(1) When the light proceeds from a luminous point 3″ from the vertical plane of projection, 6″ above the horizontal plane, and 6″ from the axis, produced, of the pyramid.

(2) When the rays are parallel and their projections on both planes make angles of 45° with the ground line.

8. Draw a bolt 4″ long, and $1\frac{1}{2}$″ diameter, with a hexagonal head, greatest diameter $2\frac{3}{4}$″, depth $1\frac{1}{2}$″. Find the shadow cast by the head on the bolt, and by both on the horizontal plane when the bolt is vertical with head uppermost. Direction of light the same as in the second case of the last example.

9. A cylinder, 2″ diameter and 3″ high, stands on a table, and a sphere of 3″ diameter rests on the top of it, with its centre in line with the axis of the cylinder. Draw the shadow cast by the two solids on the table when the rays are parallel to one another, and inclined at 45° to the table.

10. A cone of revolution, base 2″ diameter, height 4″, stands on a table; a sphere of 2″ diameter rests on the same table, the distance between the axis of the cone and the centre of the sphere being 3″. Find the shadow cast by the cone on the sphere when the rays are parallel to the plane containing the axis of the cone and centre of the sphere, and inclined at 45° to the table.

CHAPTER VI.

In most structures and machines there are three principal directions, one vertical and two horizontal and at right angles to one another. In making drawings it is the custom to take each plane of projection parallel to two of these directions. The projections on these planes are both the easiest to make and the most useful for some purposes, such as working drawings. There is, however, a considerable amount of experience and training required in order to be able readily to comprehend them; that is, to be able to combine the plan, elevation and sections so as to form to the mind a clear mental image of the object represented. When it is desired to convey a clear and easily comprehended notion of an object, especially if it be of a complicated form, to those who are not familiar with the ordinary drawings, and even sometimes to those that are, it is necessary to show the three principal directions at one view; for this has the effect of giving to the drawing an appearance of solidity. The best kind of drawing for this purpose would no doubt be a perspective; but it is of a nature to require much labour in the execution, even were it more generally understood by draughtsmen than it is. But the place of a perspective may be very well supplied in many cases by a kind of orthogonal projection on a plane inclined to the three principal directions. Three lines are taken at right angles to one another to represent the three principal directions, and are projected on a plane inclined to each of them. These lines are called the *axes*, and by measurements along and parallel to them the projection of every other point of the object to be represented may be obtained. This is what is to be understood by the term *Axometric Projection*.

From the following problem and the succeeding examples it is hoped that the method of drawing such a projection will be clear.

PROBLEM I. Fig. 108.

Given the projections of three straight lines which meet at a point and are perpendicular to one another, on a plane oblique to each of them, to determine their respective inclinations to that plane.

Let the three lines, AB, AC, AD be at right angles to one another, and let aB, aC, aD be their projections on a plane, which for convenience may be considered horizontal; it is required to find the inclination of each of them to the plane of projection.

Construction. Take any point B on one of the lines for its horizontal trace; draw BC perpendicular to aD, and CD perpendicular to aB. The points C and D are the horizontal traces of the other two lines.

Next take a vertical plane of projection parallel to one of the given lines: in the figure it is parallel to AB. That is, draw $e'b'$ parallel to aB for a ground line. On $e'b'$, the vertical projection of EB, describe a semicircle, and draw aa' perpendicular to $e'b'$ meeting the semicircle in a'; this is the vertical projection of the point A.

The vertical projections of the given lines being now known their inclinations are determined by Prob. VI. Chap. II. $a'b'e'$ is the inclination of AB, and $a'c_1'e'$, $a'd_1'e'$ the inclinations of AC and AD respectively.

Proof. Because the plane BAC is perpendicular to AD, the line BC, through the horizontal trace of AB and perpendicular to aD, is the horizontal trace of that plane (Prob. XV. Chap. II.). Therefore C is the horizontal trace of AC. For a similar reason D is the horizontal trace of AD.

Since AB is perpendicular to the plane CAD the angle BAE is a right angle; and as the vertical plane of projection is parallel to the plane BAE the vertical projection of the angle BAE must also be a right angle (Theor. XVII. Chap. I.) ; so that the vertical projection of A must be on the semicircle $b'a'e'$ and is therefore the point a'.

Hence $a'b'$ is the vertical projection of AB, and $a'e'$ the

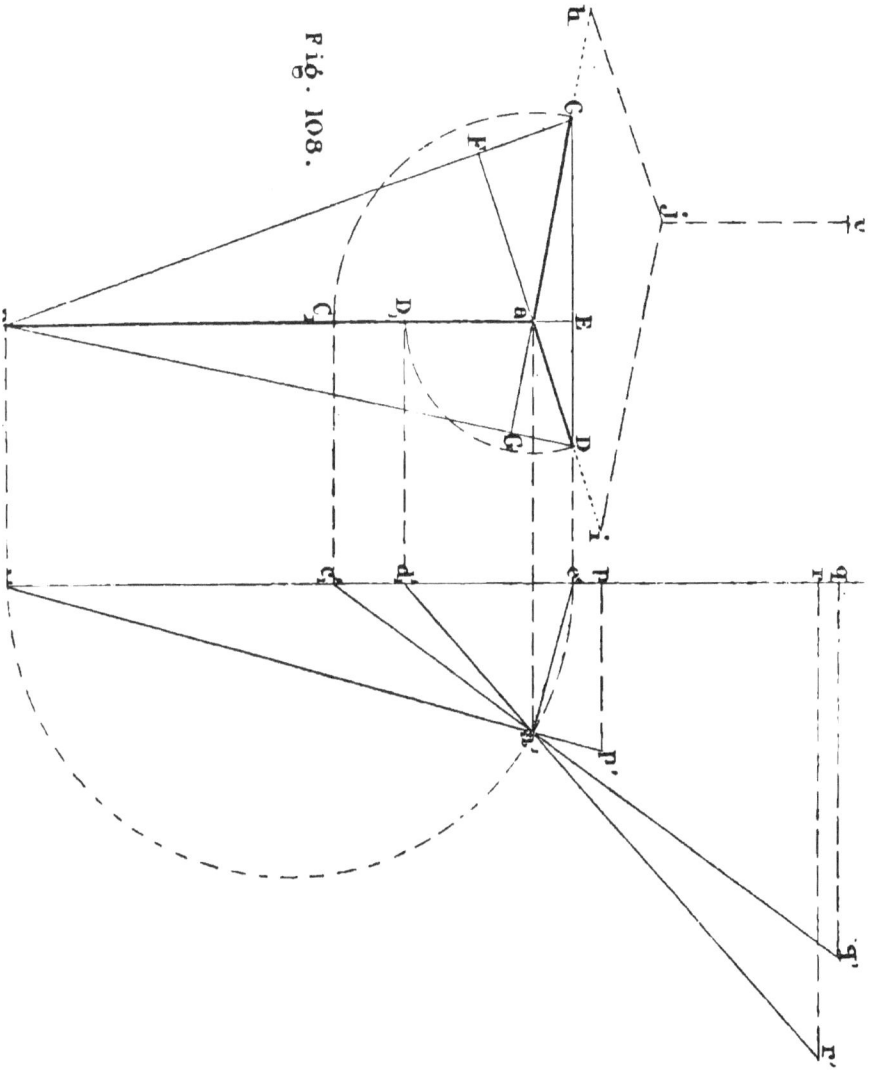

Fig. 108.

vertical projection of AC and AD. $a'b'e'$ is the inclination of AB, $a'c_1'e'$ the inclination of AC, $a'd_1'e'$ the inclination of AD...... Prob. VI. Chap. II.

Notes. When the inclinations of the lines are known the projections of any lengths measured on them can readily be found. Thus $b'p$, $c_1'q$, $d_1'r$ are the lengths of the projections of three inches on the lines aB, aC and aD respectively.

When the position of a point with respect to the planes containing the three lines AB, AC and AD is known the projection of the point may be determined. Thus to find the projection of a point V : ah is the projection of its distance from the plane BAD ; ai, the projection of its distance from the plane BAC; and jv, the projection of its distance from the plane DAC. This line jv is parallel to aB, and consequently its length is found by setting off on $a'b'$ the actual distance of V from CAD, and finding the projection of that distance on $e'b'$.

EXAMPLE I. Fig. 109.

To draw the projection of a cube of one inch side when the angles between the projections of its edges are given.

om, on and ol are the directions of the three edges. In this and the following examples the axes are taken parallel to aB, aC and aD in fig. 108, so that their inclinations are the same as in that figure.

Construction. Set off one inch on each of the lines $a'b'$, $a'c_1'$, $a'd_1'$, and find the lengths of the projections on the line $b'q$. These projections are the distances ol, om and on, which being set off on the axes from o give the angular points of the cube, l, m, and n. Each of the other edges is parallel to one of these ; so that the projection is completed by drawing through m the lines ma and mc parallel respectively to on and ol; through n drawing the lines na and nb parallel to om and ol respectively ; finally, lines being drawn through a, b, c parallel to the given edges, the figure is completed.

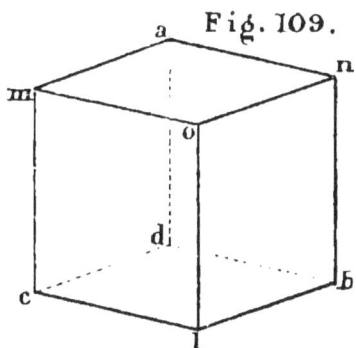

Fig. 109.

EXAMPLE II.   Figs. 110 and 111.

Figure 110 is the plan and elevation of a mortice and tenon joint; figure 111 is the axometric projection of the same: in the latter the two pieces are shown separate, the tenon being supposed to be drawn out of the mortice. The three edges of the solid through O are taken for axes. $o_1'b'$ being set off on $a'b'$, fig. 108, and its projection on $e'b'$ found gives op; pd'' is drawn parallel to on; and pb'', pd'' are made equal respectively to the projections of o_1b, o_1d, as measured on the line $a'd_1'$; $b''a''$ is the projection of ba, obtained from $a'c_1'$; $b''f''$ is the projection of $b'f'$, which is parallel to ol. The remaining lines are found by drawing through p, a'', b'', d'' and f'' lines parallel to the axes as in the last example. The rest of the construction will be evident from the figure.

Fig. 110

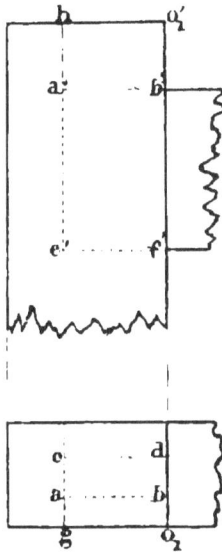

Fig. 111.

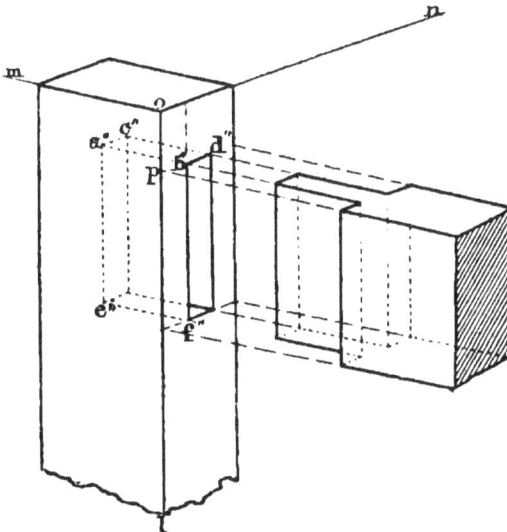

EXAMPLE III. Figs. 112 and 113.

Fig. 112 is the plan and elevation of a brass bush for the small end of a connecting rod, and fig. 113 is its axometric projection. The lines chosen for axes are the tangents to the circle $A_1B_1C_1$ at the points B_1 and C_1, and the line through O_1 at right angles to the plane of that circle. ol, om and on are the projections of these axes. Instead of finding the projection of each line of the solid directly from fig. 108 a scale is drawn for each axis and lines parallel to it. Scale (1) is that for fig. 112 as well as those lines on fig. 113 which are parallel to the plane of projection. (2) is the scale for the axis ol and lines parallel to it. It is obtained as follows :—A certain number of inches—in this case six— measured on scale (1) are set off on $b'a'$, fig. 108, from b' to p', and $b'p$, the projection of $b'p'$, represents 6 inches on scale (2). Scales (3) and (4) are those for om and on respectively, and for lines parallel to them. They are obtained in a similar way to (2) by setting off $c_1'q'$ and $d_1'r'$, each equal to 6 inches, and finding their projections $c_1'q$ and $d_1'r$.

To find the projection of the circle $A_1B_1C_1$. This may be done by any one of the three following methods.

1. The projections of any number of points on the circle may be determined by taking their respective distances from the two axes $o_1'l_1'$ and $o_1'm_1'$ and finding the projections of those distances on fig. 108 as in the two previous examples : the distances on fig. 112 being measured on scale (1), the corresponding distances on fig. 113 are found from scales (2) and (3).

2. A square being drawn circumscribing the circle, the projection of that square is the parallelogram $ompl$; that is, $ol = 4\frac{3}{4}$ inches measured from scale (2), and om the same distance measured from scale (3). The ellipse required can now be inscribed in the parallelogram $ompl$ by means of Plane Geometry.

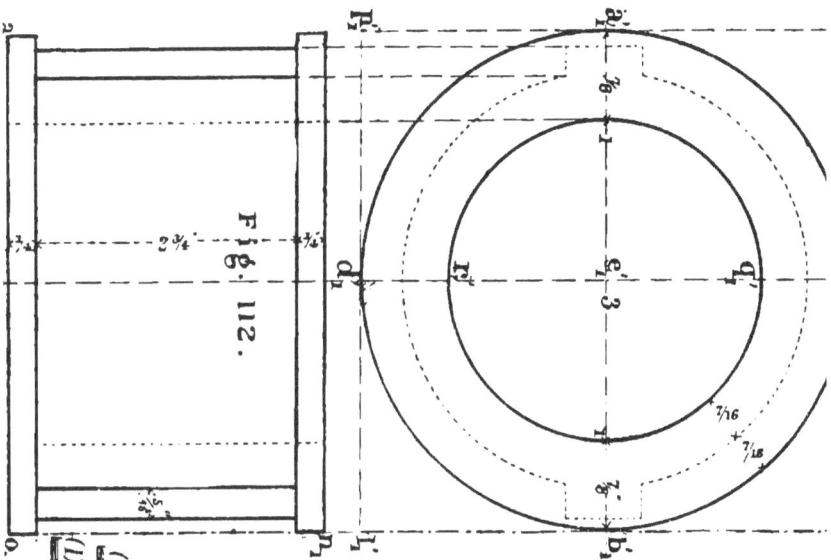

Fig. 112.

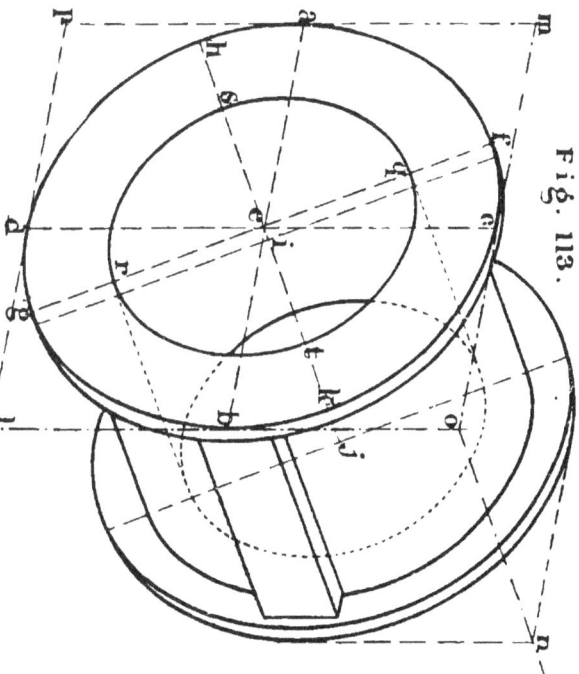

Fig. 113.

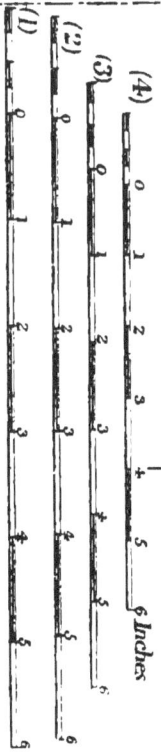

(1)

(2)

(3)

(4)

6 Inches

3. By finding the centre and axes of the ellipse. The centre of the ellipse is the projection of the centre of the circle, and is found by making $ob = 2\frac{3}{4}$ inches, measured from scale (2), and be, parallel to om, $= 2\frac{3}{4}$ inches, from scale (3). The major axis is the projection of that diameter of the circle which is parallel to the plane of projection, that is a line through e parallel to the horizontal trace of the plane of the axes ol and om. But the trace of that plane is at right angles to on. Therefore drawing through e the line fg at right angles to on, and making ef, eg each equal to $2\frac{3}{4}$ inches, on scale (1), fg is the major axis of the ellipse. The minor axis is found from fig. 108. It is the projection of a line $4\frac{3}{4}$ inches long in the plane lom, at right angles to the trace of that plane. Hence finding the inclination of AF, fig. 108, and the projection of a line of the required length, measured on it from scale (1), gives the minor axis, hk. The ellipse can now be drawn from the axes by any of the well-known methods.

The ellipse sqt has the same centre e and its axes are in the same ratio to one another; so that making eq and er each $1\frac{1}{2}$ inches, measured from scale (1), and drawing through q a line qt parallel to fk, the axes of the ellipse are found.

All the other ellipses of the Axometric Projection may be drawn in a similar manner. Their planes are all parallel to the plane of $A_1B_1C_1$, and their axes are consequently parallel to fg and hk respectively, and are in the same ratio to each other.

ISOMETRIC PROJECTION.

If two of the angles in fig. 108 were made equal to each other the two lines which contain the third angle would have the same inclination, and the same scale would in consequence serve for both. Thus if BaC and BaD were equal, aC and aD would be equal. For in the two right-angled triangles BaF, BaG, the angles BaF and BaG being equal, and Ba common to the two triangles, aF would be equal to aG; and in the two right-angled triangles CaF, DaG, the sides aF, aG being equal, and also the angles CaF and DaG, the side aC would be equal to aD. As can be readily seen from the figure, aC and aD being equal must be equally inclined to the horizontal plane.

Fig. 114.

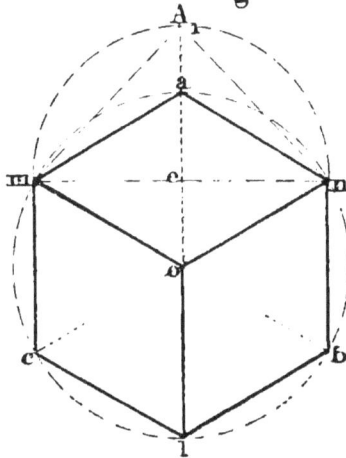

Again, if the three angles were equal to one another the three axes would have the same inclination. When the same scale serves for the three axes, the projection is said to be *Isometric*.

Fig. 114 is the isometric projection of a cube of 1 inch side. The three edges of the cube are taken for axes. These lines being drawn making angles of 120° with one another, the projection of 1 inch measured on each may be found either as in fig. 108 or as follows :—

The line *mn* being at right angles to *ol* is parallel to the plane of projection, and is therefore equal to the real length of the diagonal of the square face of the cube. If then eA_1 be drawn perpendicular to *en* and made equal to it, A_1n is the side of the square. But it is evident the angle $A_1an = 120°$ and the angle $eA_1n = 45°$; so that if a triangle be constructed, in any convenient position, having an angle of 120° and an angle of 45°, the ratio between the true length of any line parallel to one of the axes and its isometric projection is as the side opposite the angle of 120° to the side opposite the angle of 45°. In this way an isometric scale can at once be constructed from the natural scale. The method of drawing an isometric projection is so similar to that already described for any three axes that it is unnecessary to enter into further details. It is in fact a simple case of Axometric Projection.